B|
JAVA,
SUMATRA and
BALI

Tony Tilford and Alain Compost

BLOOMSBURY
LONDON · OXFORD · NEW YORK · NEW DELHI · SYDNEY

POCKET PHOTO GUIDE

Bloomsbury Natural History
An imprint of Bloomsbury Publishing Plc

50 Bedford Square
London
WC1B 3DP
UK

1385 Broadway
New York
NY 10018
USA

www.bloomsbury.com

BLOOMSBURY and the Diana logo are trademarks of
Bloomsbury Publishing Plc

First published by New Holland UK Ltd, 2000 as *A Photographic Guide to
Birds of Java, Sumatra and Bali*
This edition first published by Bloomsbury, 2017

British Library Cataloguing-in-Publication Data
A catalogue record for this book is available from the British Library.

Library of Congress Cataloguing-in-Publication data has been applied for.

ISBN: PB: 978-1-4729-3818-3
ePDF: 978-1-4729-3816-9
ePub: 978-1-4729-3819-0

2 4 6 8 10 9 7 5 3 1

Designed and typeset in UK by Susan McIntyre
Printed in China

To find out more about our authors and books visit www.bloomsbury.com.
Here you will find extracts, author interviews, details of forthcoming events
and the option to sign up for our newsletters.

CONTENTS

INTRODUCTION

The islands of Java, Sumatra and Bali total some 613,000 square kilometres, almost 2½ times the area of the United Kingdom, and possess approximately 635 bird species. With such a rich avifauna, it is scarcely surprising that birds are widely kept there for their beauty and song. Visitors, who now reach the region in ever-increasing numbers, cannot fail to notice that almost every house has caged birds hanging from the eaves, often in large numbers. Close inspection reveals that many of the birds kept in captivity here are often the same as those to be found in pet stores around the world. This is a reflection of the fact that many birds traded on the international pet market are local species from this region.

Trapped and exported in increasingly greater numbers, natural populations of many species are severely reduced, and some have been brought to the edge of extinction in spite of protective legislation. So far, only the Javanese Lapwing has become extinct in recent times, but several other species, most notably the Bali Myna, are hovering on the brink. With fewer than 30 individuals of this striking species left in the wild in western Bali, in 1999, it is no coincidence that some 3,000 are to be found in aviaries throughout the world. The demise of the Bali Myna is well documented and is largely the result of illicit trapping, exacerbated by human disturbance and loss of habitat. The same dismal picture is repeated for many other species, including the Java Sparrow, Brahminy Kite and Black-winged Starling, all of which were, not many years ago, a common sight.

It is fortunate that the Indonesian authorities, in particular the Directorate General of Forest Protection & Nature Conservation of the Indonesian Ministry of Forestry (PHPA) in conjunction with BirdLife International's Indonesia Programme, have recognised the urgent need for protection, and vast areas have now been designated as National Parks and reserves with special protection for wildlife. Visitors should bear in mind that permits are officially required to enter many of these reserves and prior arrangements may be necessary. Even outside the reserve areas the most threatened animals are protected, at least in theory. For further information, the World Wide Web addresses given later in this book are a good place to start.

Within the space limitations of this small book it would be impossible to cover all the 635 or so bird species that have been recorded in Java, Sumatra and Bali. The selection has therefore been based upon the birds that are reasonably common throughout the region. A few less common but interesting and more spectacular species have been added, so as to reveal the enormous biological diversity to be found there and to provide an understanding of its characteristic nature.

The 236 species treated include a reasonable representation from most of the families likely to be encountered in the region. The reader will undoubtedly come across others, and these are certain to be covered in the more specialized (but unfortunately more cumbersome) books listed for further reading.

BIRDWATCHING

For most of us, birdwatching is really putting a name to a bird, but there is so much more to be gained from more detailed study. Behaviour, life cycles and interaction with the environment are fascinating subjects which not only can give us enormous enjoyment, but can add much to our knowledge of the world around us.

Initially, however, identification must be the first priority. We should consider all the clues available and, ideally, identify not just the species but also its sex, age and race where possible. There are always many clues to go on, and by combining them all we narrow down the possibilities until our problem is solved. We all accept the value of visual clues, but tend to neglect the importance of sound until we become more experienced. For the seasoned birdwatcher, his or her tape recorder is just as important a part of the kit as are note-book, binoculars, telescope and field guide. That important call can often be identified later in the day. It is always a help to have a checklist of the species found in the area, and a good place to find this is on the World Wide Web. Try some of the addresses listed under 'Useful Web-sites' later in the book; it is amazing what information is available.

Whatever equipment you take, it should be well protected from the ravages of the weather. Rain and high humidity provide the worst conditions for cameras, binoculars and electronics. Any equipment should be kept dry and well aired and, if at all possible, stored with a bag of silica-gel desiccant to protect against fungal attack. It is always astonishing how much damage fungus can cause if inadequate precautions have been taken.

Ideally, lightweight waterproof binoculars such as the Nikon close-focusing WP/RA II range are recommended, particularly if you are going to venture into wet and humid areas, but for general use the not too expensive kind, such as the Nikon Travelite IV range, is perfectly adequate for most observation. For the more serious birder there are many other 'professional-style' binoculars to choose from, as well as spotting scopes for more distant observation. For sound recording, the Sony Mini-Disk Recording Walkman with a directional microphone gives superb reproduction, but for identification purposes many of the smaller cassette recorders are equally useful. As for cameras, they are not recommended for regular birdwatching trips, not only because of the possibly detrimental effect of the weather on equipment, but also because it is almost impossible to concentrate on both activities at once, particularly in a group.

It is fortunate that many of the good birdwatching areas are accessible by car, but there is always public transport and, except in Bali, at the end of the journey the ubiquitous motorbike-taxis known as 'ojeks'. In the field, the energy-sapping heat and humidity are perhaps the biggest drawbacks. It should be unnecessary to advise here on hygiene and the need to carry sufficient bottled water. Keeping fit and healthy is extremely important, particularly in areas where medical facilities may be poor or non-existent. Do not forget to take essential medication with you, and always be prepared for stomach upsets. Remember, 'Bali Belly' is not confined just to Bali. Insects can also be a problem at certain times, and insecticide sprays and creams could save you from some nasty bites. Suitable clothing is also essential, not only as protection against the elements but also as a disguise. Lightweight, dull-coloured clothes are really necessary. Visitors are recommended to carry a light poncho with a hood, and in lowland forests long trousers are very desirable for preventing leeches and stinging plants from making contact with your skin. Sturdy footwear is a must, although many people prefer heavy-duty sandals of the 'Teva' type, particularly in wet conditions. Lightweight canvas boots are better than leather ones, which quickly become mouldy.

HOW TO USE THIS BOOK

In line with other books in this series, this one uses symbols as a guide to families and family groups to which the birds belong; a key to these symbols is shown opposite. Photographs depict the bird in its commonly seen form where possible, and variations are well described in the text. The order of the 236 species covered generally follows the Peters sequence adopted by Paul Andrews in *The Birds of Indonesia: A checklist*. Vernacular and scientific names are generally those used by T. Inskipp, N. Lindsey and W. Duckworth in *An Annotated Checklist of the Birds of the Oriental Region*. Other checklists exist, and the process of revision based on more recent scientific knowledge will continue. It must therefore be expected that not only will common names vary, but so too will the scientific. Ultimately, the scientific names prove more reliable.

Each species description includes a common name followed by its scientific name, and then the species' overall length from the tip of the bill to the end of the tail. The remaining description follows no particular order but provides most of the clues necessary for a fairly reliable identification. Only for a few species has any attempt been made to describe calls and song, as it is felt that written descriptions lack meaning without the experience of hearing the actual sound. Nevertheless, the value of sound must not be underestimated. In many cases, it can be the only determining factor in successful identification. Much of the description is necessarily brief and written in a semi-technical language adopted by most bird books, but for easy understanding many of the technical terms are explained in the following glossary, and the illustration showing bird topography (below).

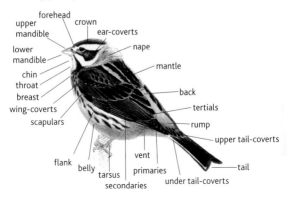

6

KEY TO COLOURED TABS

Petrels to pelicans

Herons, egrets & storks

Raptors

Ducks & geese

Gamebirds

Rails & allies

Waders

Terns

Pigeons & doves

Parrots

Cuckoos & allies

Owls & nightjars

Swifts

Kingfishers & bee-eaters

Dollarbird & hornbills

Barbets & woodpeckers

Broadbills & pittas

Larks, swallows, wagtails & pipits

Minivets

Bulbuls

Leafbirds

Shrikes & shortwings

Robins & chats

Thrushes & babblers

Mesias, prinias & warblers

Flycatchers, fantails & whistlers

Tits & nuthatches

Flowerpeckers, sunbirds & spiderhunters

White-eyes

Avadavats & munias

Sparrows & weavers

Starlings, mynas & orioles

Drongos & woodswallows

Jays & crows

GLOSSARY

axillaries Feathers of the `armpit'

canopy Unbroken layer of branches and foliage at tops of trees in forest

cere Bare fleshy or waxy protuberance at base of upper mandible, including nostrils

chin-stripe A marking around area of chin

coverts The small feathers at base of quill feathers forming main flight surfaces of wing and tail

dimorphic Occurring in two genetically determined plumage forms

echolocation Navigation by ultra-sound radar

endemic Indigenous species restricted to a particular area

eyebrow Contrasting line above eye (supercilium)

eye-stripe Contrasting line through eye

flank Side of the body

frontal shield Skin or hard unfeathered area on forehead which extends to bill

fulvous Pale brownish-yellow

Greater Sundas Borneo, Sumatra, Java and Bali, including offshore islands

gregarious Frequently occurring in groups

gular pouch Bare, fleshy patch of skin around neck of cormorants and hornbills

gunung Indonesian name for mountain

hackles Long, narrow and often pendulous feathers around neck

lores Area of feathers between bill and eye

mandible Upper or lower half of bill

malar In area between base of bill and side of throat

mesial Dividing down the middle

migrant Non-resident traveller

montane Relating to mountain habitats usually above 900m

necklace A line of markings around front of neck

Nusa Tenggara Area covered by the islands of Lombok, Sumbawa, Sumba, Flores and Timor

orbital ring Unfeathered bare ring around eye

primaries The main outer flight feathers (show as longest part of folded wing)

primary forest Original natural forest

race Another name for subspecies

rackets Paddle-shaped ends to tail feathers

resident Remaining in a local area throughout year

roost Resting or sleeping place

secondaries The inner flight feathers on rear half of wing

secondary forest New forest replacing primary forest

spatulate Having thickened rounded ends

speculum Contrasting iridescent patch on a duck's wing

subspecies A population which is morphologically different from other populations of same species

Sundas Region covered by the Greater and Lesser Sundas and Sulawesi

supercilium A stripe above eye (eyebrow)

terminal At the end or tip

underparts Undersurface of body from throat to undertail-coverts

undertail-coverts Small feathers below tail covering bases of tail
   feathers
underwing-coverts Small underwing feathers covering bases of
   primaries and secondaries
upperparts Upper surface of body
vent Area around anus, including undertail-coverts
wattles Brightly coloured bare skin hanging from head or neck
wingbar A visible line of colour at tips of wing-coverts
wing-coverts Small feathers on wing covering bases of primaries and
   secondaries

THE REGION

Java, Sumatra and Bali lie along the edge of the South-east Asian
tectonic plate, where it rides over the Australian-Indian plate
generating intense geological activity. A line of volcanoes, some
still active, others dormant, forms a spine along the island chain. In
Sumatra these create the Barisan mountain range, stretching the
entire length of the island, and falling precipitously into the depths of
the Indian Ocean in the west; in the east it drops more gently, giving
rise to large swampy areas before reaching the Sunda Sea. In Java,
the mountains arise in isolation from alluvial plains and, as in east
Sumatra, slope gently northwards to the Sunda Sea.

At one time Bali was connected by landbridge to Java, but today
it is separated by some 3km of turbulent waters, renowned for their
fierce currents. Not surprisingly, the fauna and flora of Bali show close
similarities to those of eastern Java. Sumatra, however, shows closer
affinities with Borneo than with Java.

The fragmentation of the landmasses combined with geological
and climatic activity has resulted in the formation of many habitats,
which in turn support a rich diversity of wildlife. This includes many
endemic species: those confined to a single region or smaller area and
found nowhere else.

Java and Bali are among the most densely populated regions on
earth, and the destruction of their remaining wildlife habitats is a
matter of serious concern. Only on remote mountain slopes can one
find significant areas of natural forest. A few patches of lowland forest
survive in National Parks and Nature Reserves. Sumatra, with only a
tenth of the population density, is a little better off, but the growth
of the agriculture, plantation and forestry sectors are responsible for
increasing forest destruction there.

THE AVIAN FAUNA

We might expect that these landmasses, which were at various times
joined to mainland Asia, would have fairly similar flora and fauna. As
the various islands were not all separated simultaneously, however,
and since subsequent local climatic conditions have varied, the
composition of the avian fauna differs significantly from one island to
another. Perhaps the most dramatic changes have been caused more
recently, by man and his destruction of the lowland forests. There are
ample records dating back as far as 1657, and it is quite clear from the

reports of the early 18th-century explorers who founded Indonesian ornithology that birds were far more plentiful then.

WHEN AND WHERE TO FIND BIRDS

As the region lies close to the equator, the climate is hot and humid. The monsoon weather patterns of the region are characterized by wet and dry seasons, which differ little in temperature. Even so, the birds show distinct breeding cycles. During the wet season, from October to March, large numbers of wintering migrants fly down from the north, with a smaller number of Australian species appearing during the dry season.

Many waterbirds arrive at the end of the wet season, nesting in the safety of isolated trees in a flooded landscape. This is also the peak of the insect breeding season, when abundant food is available for insectivorous species to rear their young. Frugivorous species delay breeding a little longer, until the trees and bushes are bearing ripe fruit.

Good birding areas, with a rich diversity of habitats, abound throughout the region. Clearly, it is not possible to list them all in a book of this size, but, to guide the birdwatcher to some of the more productive areas, the major reserves and National Parks are described below. These are shown on the accompanying map.

IMPORTANT BIRDWATCHING SITES

SUMATRA:

Gunung Leuser National Park

This is a huge area of rainforest and mountains with a small coastal and lowland extension. It is an ideal place to explore the birds of northern Sumatra, such as Wreathed, Black, Oriental (or Asian) Pied, Helmeted and Rhinoceros Hornbills, Great Argus, Crestless Fireback and Crested Partridge, Brown, Yellow-crowned and Gold-whiskered Barbets, White-rumped Shama and Black-and-yellow, Green and Dusky Broadbills, Crested Jay and Red-bearded Bee-eater.

Kerinci-Seblat National Park

The park includes the dominating peak of Mt Kerinci and the Kerinci valley wetlands. The best place to see Sumatran montane birds, as well as several of the endemics such as Schneider's Pitta, Bronze-tailed Peacock Pheasant and Sumatran Cochoa. Also recorded there are Salvadori's Pheasant, Silver-eared Mesia, Chestnut-capped Laughingthrush, White-throated Fantail, Sunda Bush Warbler and Golden Babbler.

Way Kambas National Park

Situated in the south-east of Sumatra, Way Kambas includes remnants of lowland rainforest and coastal and swamp forest. A large number of species have been recorded there, including White-winged Duck, Storm's Stork, Stork-billed Kingfisher, Lesser Adjutant Crested Fireback, Crested Partridge, Great Argus, Hill Myna, Black-bellied Malkoha and Scarlet-rumped Trogon, Gould's, Sunda and Large Frogmouths, Black-thighed Falconet and Orange-breasted and Cinnamon-headed Green Pigeons.

Berbak National Park

This reserve is on the east coast and includes coastal forest and peat swamp, as well as mangroves. It is a good area for Büttikofer's Babbler, Wallace's Hawk Eagle and Milky and Storm's Storks.

Bukit Barisan Selatan National Park

Situated at the southern tip of Sumatra, at the end of the Barisan mountain range, the park reaches from the sea to the top of Gunung Pulung and includes all forest types and some very wild areas. It is relatively unrecorded but could bring some interesting surprises. Among the rarer birds in the area are Sumatran Treepie, Helmeted Hornbill, Red-billed Partridge, Lesser Adjutant and Milky and Storm's Storks.

JAVA:

Ujung Kulon National Park

At the extreme west tip of Java, Ujung Kulon boasts many rare species in a great variety of habitats, from lowland rainforest and evergreen forest to coastal scrub, mangroves and open grazing areas. As many as 24 endemic or threatened species are to be found there, including Javan Coucal, White-breasted Babbler, Javan Sunbird, Javan Hawk Eagle, Blue-throated Bee-eater and Green Peafowl.

Gede/Pangrango National Park

This West Javan National Park consists mainly of luxuriant evergreen submontane forests, but with mossy elfin forest and alpine meadows at the peak of Gunung Pangrango. It is a wonderful place to see many of the Javan endemics, such as Javan Tesia, Javan Cochoa, Javan Hawk Eagle, Chestnut-bellied Partridge, Javan Scops Owl, Javan White-eye, Dusky (Horsfield's) Woodcock, Pygmy Tit, Mountain Serin, Volcano Swiftlet and the Blue-tailed Trogon.

Meru Betiri National Park

Meru Betiri covers large areas of moist primary and secondary forests, mangroves and a rather rugged coastline with some beautiful beaches. It is one of the less explored areas where new discoveries are likely to be made. Among the rarer birds recorded are Wreathed Hornbill, Banded Woodpecker, Violet Cuckoo, Pin-tailed Parrotfinch, Black-crested Bulbul, Grey-cheeked Green Pigeon, Black-naped Fruit Dove, Javan Owlet, Waterfall (or Giant) Swiftlet, Black-banded Barbet and Crescent-chested Babbler.

Baluran National Park

Situated at the north-east end of Java, Baluran National Park is spread around the dormant volcano of Gunung Baluran. It is one of the easier places for birdwatching, being more open and easily accessible. A good place to see Green Peafowl and Green and Red Junglefowl. White-bellied Woodpecker, Oriental Pied Hornbill, Spotted Wood Owl, Banded Pitta, Lesser Adjutant and Grey-cheeked Tit Babbler are all reported from the area.

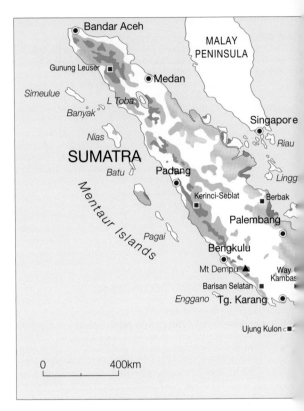

Kebun Raya (Botanic Gardens), Bogor

Being easily accessible, Bogor's Botanic Gardens become crowded with local visitors at the weekend. A weekday visit, however, can be very fruitful in the comparative tranquillity. The enchanting settings of Kebun Raya, covering a 1-kilometre square, have been established for around two centuries and provide some very ready opportunities to see many town and woodland birds. There is a roost of Black-crowned Night Herons, and it is difficult to miss the Black-naped Orioles, Sooty-headed Bulbuls and Oriental Magpie Robins. Other species likely to be seen are Black-naped Fruit Dove, Collared Kingfisher, Blue-eared Kingfisher, Grey-cheeked Green Pigeon, Horsfield's Babbler, Olive-backed Sunbird, Purple-throated Sunbirds, Hill Blue Flycatcher and the Yellow-throated Hanging Parrot.

Pangandaran Nature Recreation Park

Just west of Cilacap on the south coast of Central Java is the peninsula of Pangandaran, taken up by the Nature Recreation Park. Consisting mainly of dry forest, it is a good area for Green Junglefowl and Banded Pitta. Other species present are Oriental Pied Hornbill, White-rumped Shama, Scaly-crowned Babbler and Black-backed Kingfisher. To the

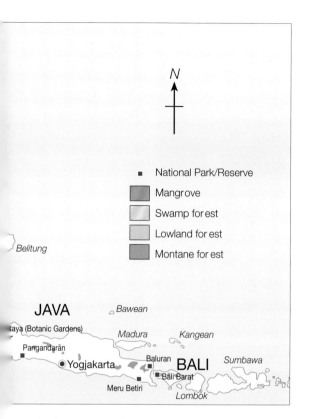

east lie the mangroves of Segara Anakan, where many waterbirds can be seen. Among the large population of herons, egrets and terns, Lesser Adjutant and Milky Stork may be seen. Javan Coucal and Stork-billed Kingfisher have also been recorded there.

BALI:

Bali Barat National Park
Situated at the western tip of Bali, Bali Barat is a mixture of dry and moist forests with coastal scrub, savanna and mangroves. It is the only place to find the wild, endemic Bali Myna. Many other interesting species reside there, including Banded Pitta, Green Junglefowl, Pink-headed Fruit Dove, Black-backed Fruit Dove, Shiny Whistling Thrush, Sunda Bush Warbler and Lemon-bellied White-eye.

VISITING NATIONAL PARKS

Permits to enter national parks can be obtained at the parks' entrances. In most Indonesian national parks, facilities such as trail networks, the provision of experienced bird guides, and accommodation close to the best birding sites, are limited.

WILSON'S STORM-PETREL *Oceanites oceanicus* 17cm

Roger Tidman, Windrush Photos

This diminutive species characteristically flies low over the sea, making slow progress as it rises and dips, moving from side to side searching the water surface much like a butterfly. In flight, its general plumage appears blackish-brown with a paler wingbar and a very conspicuous white rump and uppertail-coverts. The tail also appears short and square, with the feet protruding behind. It is the commonest small oceanic bird likely to be encountered on the coast. It arrives in the region after breeding mainly on small islands in the southern oceans, and spends its time at sea feeding on floating organic matter and small crustaceans.

LITTLE GREBE *Tachybaptus ruficollis* 25cm

Ray Tipper

This tiny waterbird appears as a very small dark duck swimming high in the water, repeatedly diving and remaining submerged for prolonged periods. During the breeding season, its upperparts are brown, becoming darker on the crown and nape, while the throat and the sides of the neck are chestnut. It has an obvious yellow patch at the base of the black bill. In non-breeding plumage it is greyer, the neck becoming whitish. It is a rare resident in Java, Bali and northern Sumatra, where it frequents lakes and flooded rice paddies, although its range extends from New Guinea to Europe.

WHITE-TAILED TROPICBIRD *Phaethon lepturus* 39cm

Tilford/Cooke, TC Nature

Adult birds are usually white, some with a yellowish tinge, and have long white tail-streamers. They are marked with black on the wing-tips and have a black bar on the upperwing. Juveniles have no tail-streamers, and their upperparts are barred black. Although resident in some parts of the region, visitors also arrive from other areas. Individuals occurring around Sumatra, and recognizable by their golden plumage, belong to the subspecies (*P. l. fulvus*) which comes from Christmas Island in the Indian Ocean. Breeding colonies have been located on Nusa Penida, Bali, and along Java's south coast.

LITTLE BLACK CORMORANT *Phalacrocorax sulcirostris* 60cm

A bird with fairly uniform black plumage with a purplish-green iridescence, except for mottling on the wing-coverts. In the breeding season, it acquires a small patch of white feathers behind the eye. It has a small greyish gular pouch and bare skin around the eye. The bill is grey-blue and the feet black. Mainly a bird of lakes, estuaries and fishponds, it is occasionally seen along the coast. This species' range extends from Australia up through New Guinea and westwards to the Greater Sundas, but few birds reach Sumatra. It is not a common bird, although the number present at Pulau Rambut on the north-west Javan coast might lead one to believe that it is.

David Tipling, Windrush Photos

LITTLE CORMORANT *Phalacrocorax niger* 56cm

This is the smallest of the cormorants found in the region. The plumage is generally greenish-black, with a few tiny white feathers on the neck and head. The bill is brown, with a black tip and purple near the base. After breeding, the plumage darkens and loses the white flecks, except for the chin and throat, which become white. It frequents mangroves, marshland, lakes and estuaries, often in small flocks swimming low in the water and diving for fish. A fairly common resident of the coasts and lowland waterways of Java. Those seen in Sumatra are likely to be visitors from Java.

Chew Yen Fook

ORIENTAL DARTER *Anhinga melanogaster* 84cm

Very similar to the cormorants apart from the long slender neck, small head and yellow-brown dagger-like bill. It is a mainly black bird, with white-streaked plumed coverts on the wing and back. It swims very low in the water with only its head and neck showing, so it often becomes waterlogged and has great difficulty in taking to the air. It spends a lot of time standing alone on low, exposed perches, often with its wings outstretched to dry, but it roosts communally. Darters prefer to inhabit large, clean freshwater lakes and forested rivers where they dive underwater in search of fish.

Chew Yen Fook

AUSTRALIAN PELICAN *Pelecanus conspicillatus* 150cm

Very similar in size to the Great White Pelican *Pelecanus onocrotalus*, the Australian Pelican can be identified by its black tail and upper-wing coverts, pinkish gular pouch stretching the full length of the very large pinkish-purple bill, and lack of bare facial skin patch. The primaries and secondaries are black; legs and feet are grey-blue. It flies with a laboured wing-beat and catches fish by plunging into the water. This pelican is occasionally found singly but more usually in small flocks on estuarine sand bars. It breeds in Australia, migrating north to New Guinea and west as far as Bali and E. Java.

GREY HERON *Ardea cinerea* 100cm

A predominantly grey bird with paler underparts, the Grey Heron has a prominent black streak on the head and black flight feathers. It also has a line of black streaks down the front of its neck. Legs and bill are yellowish. It appears smaller than the Purple Heron and lacks that species' reddish-brown coloration. Found mostly in the lowlands, these herons frequent wet areas such as paddyfields, lakes, mangroves and swamps, hunting fish, frogs and crabs. This is a colonial nester, often seen in tall trees in the proximity of mangroves. Widely distributed, it is resident throughout the Old World temperate regions as well as the tropics.

GREAT-BILLED HERON *Ardea sumatrana* 115cm

The largest heron found in the region, the Great-billed Heron has a dark greyish-brown plumage and shows a small crest. Its feet are grey and the bill is grey-black. It has a repetitive 'roaring' call, as well as low-pitched croaking. It prefers a habitat of mangroves and mudflats, and also frequents the beaches and coastal reefs of small islands. It is generally solitary in its habits, and may be seen stalking along the reefs and creek edges in search of small fish. Although not common, this species is a resident throughout the Greater Sundas.

PURPLE HERON *Ardea purpurea* 95cm

Chew Yen Fook

Similar in height to the Grey Heron, this more elegant, slimmer bird has a less hunched stance. It is easily distinguished from other herons of its size by its dark underparts and its black-streaked russet-coloured neck and upper breast. The back and wings are greyish-purple with a scattering of long russet plumes. It has a brown bill and reddish-brown legs. Normally a more solitary species, it congregates in large colonies to nest. It is found in typical heron habitat of paddyfields, swamps, mangroves and lakes, often inland, where it feeds on small fish and amphibians.

GREAT EGRET *Ardea alba* 90cm

The largest of the white egrets, with a more substantial bill and a peculiar kink midway along the neck. During the breeding period, the bill is black, the bare facial skin becomes greyish-blue and the bare thighs turn red. At other times of the year, the facial skin is dull yellow, the bill is yellowish, often tipped black, and the feet and legs are all black. It prefers mudflats, mangroves and coastal marshland, as well as rice paddies, where it is found either alone or in small groups, generally feeding on small fish and frogs. Although this species occurs throughout the Greater Sundas, nowhere is it common.

INTERMEDIATE EGRET *Egretta intermedia* 68cm

Chew Yen Fook

Also known as the Plumed Egret because of the adornments of its breeding plumage (although other egrets also develop such plumes to a greater or lesser degree when breeding). This all-white bird is in fact intermediate in size between the Great and Little Egrets, but it is less common than either. It may be distinguished from the smaller Little Egret by its yellow bill, and from the larger Great Egret by its smooth, unkinked neck. In common with its relatives, it can be found in paddyfields, mangroves and swamps and on coastal mudflats.

LITTLE EGRET *Egretta garzetta* 60cm

Larger and slimmer than the Cattle Egret, this species has greyish-black legs and a black bill. Normally all brilliant white in plumage, in the breeding season it develops long pendulous plumes on the breast, back and nape. Its small, dirty yellow bare facial patch becomes pink during the breeding season. The Asiatic migrant race is distinguished by having yellow toes. This common heron is found mainly in mixed flocks in coastal lowlands, where it occurs in paddyfields, on riverine mudflats and sandbars and alongside small streams. It mixes freely with other herons and egrets in their breeding colonies and at night-time roosts.

PACIFIC REEF EGRET *Egretta sacra* 58cm

The Pacific Reef Egret possesses two colour morphs. The commoner form has a uniform grey plumage, a short crest and a very pale chin. Less common is the white form, which resembles the Cattle Egret but is considerably larger, and has a sleeker head and neck. It has relatively short greenish legs and a pale yellow bill. This species is generally confined to the coastline, where it can be seen resting on rocks and cliffs when not hunting at the water's edge. It is found more commonly on the sandy beaches and reefs of offshore islands.

CATTLE EGRET *Bubulcus ibis* 50cm

Commonly found in paddyfields and freshwater swampy areas, this species is also attracted to grazing cattle, feeding on the insects they disturb. A smallish bird, it is normally white, sometimes with a faint tinge of orange on the forehead. In breeding plumage, however, it develops an orange wash over the breast, head and neck, as well as red bill, legs and lores. Its more rounded body, short neck and short, thick bill distinguish it from other egrets. It nests and roosts colonially. In Bali, a large nocturnal roost (shared with other species) in the village of Petulu is protected by the belief that the birds carry the souls of Balinese killed in the 1966 anti-Communist slaughter.

JAVAN POND HERON *Ardeola speciosa* 45cm

Seen singly or in small groups, often in the company of other waterbirds, as it slowly stalks its prey in muddy areas or at the waterside. This heron normally appears dark brown, streaked with light brown, with buffish-brown tail and underparts and white wings, but in breeding plumage the neck and head become golden-buff and the back blackish. It has a black-tipped yellow bill and dull green feet. Found in both inland and coastal areas of Java and Bali, where it is fairly common on freshwater marshes and paddyfields. In south Sumatra, it is thought to be only a non-breeding visitor.

LITTLE HERON *Butorides striata* 45cm

Also known as the Mangrove Heron or Striated Heron, this little bird is a solitary hunter which will stand motionless in one position, awaiting its prey. Adults are mainly dark grey, with a very dark greenish crown and long crest feathers; they have a pale buff neck and facial markings. Juveniles have a more stocky appearance, and are brown with a streaked breast and mottled upperparts. Quite common, particularly around the coastal wetlands, this shy bird prefers to remain close to vegetation cover, but it is infrequently seen also on rocky shorelines and exposed reefs.

BLACK-CROWNED NIGHT HERON *Nycticorax nycticorax* 60cm

Adult

Juvenile

Adults have a distinctive plumage of black back and crown, white neck and underparts, and grey wings and tail, with two long white plumes emerging from the nape. In breeding plumage, the normally dirty yellow legs become red. Females are slightly smaller than males. Juveniles are streaked and mottled brown. By day these birds congregate to rest in tree roosts, but as evening descends they circle the roost, giving characteristic croaking calls, before leaving for freshwater feeding areas in paddyfields, swamps and mangroves and on mudflats. They often breed in large and very noisy colonies in mangroves or in trees overhanging water.

MALAYAN NIGHT HERON *Gorsachius melanolophus* 50cm

As it is nocturnal and very timid, and prefers dense vegetation, particularly bamboo and reed surrounding inland marshes, this species is seldom seen except when it ventures out into open areas to feed. It is solidly built, having chestnut upperparts finely barred and speckled across the back and wings, and a pale chestnut neck and head with black trailing crest. Its underparts are white, and flecked heavily with chestnut, buff and black; the throat is white, streaked black along the median line. It has olive-green legs, and the short stubby bill is olive-brown with yellow on the lower mandible. A winter visitor to Sumatra and its islands and occasionally Java.

YELLOW BITTERN *Ixobrychus sinensis* 38cm

Chew Yen Fook

Rice paddies, reedbeds, swamps and beach forest are all favoured by this very quiet and secretive bird. It occasionally appears briefly in the open, only to dart back into the reeds and disappear. In flight, the adult's warm buffish-brown upperparts often appear yellow, contrasting with the black tail, flight feathers and crown, while the underparts are buff. Juveniles are duller, with brown streaking. Feet and bill are brownish-yellow. Although an uncommon resident in Sumatra, this is a common winter visitor in Java and Bali. Sadly, as it reaches the northern coasts of Java, it is caught in fairly large numbers to be fried and eaten with other shorebirds as 'Burung Goreng'.

CINNAMON BITTERN *Ixobrychus cinnamomeus* 41cm

One of the smaller, more timid bitterns of the region which remains under cover while stalking its prey in grassland, freshwater swamps and rice paddies. When disturbed, it rises quickly from cover with a croaking alarm call, and flies off with slow powerful wingbeats. Adult males have fairly bright cinnamon-brown upperparts and orange-buff underparts, and are streaked black down the centre of the belly and along the upper flanks; the sides of the long neck are streaked dirty white. Females and juveniles appear darker, with a black cap, but juveniles have a more mottled appearance. Although very common, it is usually found alone.

MILKY STORK *Mycteria cinerea* 90cm

This large stork normally occurs in single species groups but is occasionally found in the company of other storks and herons. It is primarily white but with black flight feathers. It has a prominent bare patch of facial skin which varies in colour from pink to red, a long decurved olive-yellow bill and greyish legs. Juvenile birds are a dirty pale brown, but with the rump white and the flight feathers black. A relatively rare species but found throughout the region, it occurs in mangroves and mudflats of West Sumatra, the northern coast of Java near Jakarta, and Brantas Delta, near Surabaya.

WOOLLY-NECKED STORK *Ciconia episcopus* 85cm

This large stork is predominantly black, but has a very 'fluffy' feathered white neck. The undertail and lower belly are also white, as are the forehead and a narrow eyebrow. It has a patch of dark grey facial skin. The feet are a dirty red, while the bill is blackish, tinged red, with a red tip. Immature birds differ from adults in having the black plumage tinged brown. Although uncommon in Bali and Java, this species can be seen feeding in paddyfields and pastureland in the company of other storks, and often roosting with them in tall trees. It is not a colonial breeder.

STORM'S STORK *Ciconia stormi* 80cm

Another of the region's very rare storks, this species is similar in appearance to the Woolly-necked Stork. It has predominantly black upperparts, a black breast, white neck with black patterning at the side, and a black crown. The tail and belly are white. During the breeding season, the bright pink facial skin patch and the yellow eye-ring are very evident. The slightly upturned bill is reddish and the legs are pink. A good place to view this bird, which is found only in the freshwater pools in lowland forest in Sumatra and West Java, is Way Kambas in Sumatra.

LESSER ADJUTANT *Leptoptilos javanicus* 110cm

Chew Yen Fook

Not a pretty bird, this is a strange-looking creature. Its massive bill on a virtually bald head, and patches of bare red skin on the head, neck and upper breast, give it the appearance of being deformed and sick. The back, wings and tail are dark grey, and the underparts and upper back are white. The bill and legs are greyish. Juveniles have brownish-grey upperparts and dirty white underparts. Hunting and habitat loss are responsible for the rarity of this bird in Java and Bali, but it remains locally common in lowland areas of south Sumatra. It is a colonial nester, and often joins with other storks and raptors to soar on thermals.

GLOSSY IBIS *Plegadis falcinellus* 60cm

The only ibis in the region with all-dark plumage, this species has blackish-chestnut coloration with green, purple and bronze iridescences. Immature birds are dark brown, with buff streaking on the head and neck. Usually found in small groups in paddyfields, marshland and lake fringes, where it probes the mud with its long bill. It often associates with egrets and herons, both at night-time roosts and at breeding colonies. A good place to see this bird is at its only Javan breeding colony at Pulau Dua. Although this ibis has an almost worldwide distribution, it is uncommon in the region.

BLACK-SHOULDERED KITE *Elanus caeruleus* 30cm

At first sight, particularly in flight, this elegant little raptor might be mistaken for a gull. Its grey upperparts and white underparts, along with long black primaries and fairly squarish tail, can be confusing. Its distinguishing features are the black shoulder patches and the fact that it often hovers like a kestrel before dropping on its prey. It prefers open countryside and can be found in dry areas of sparse woodland, along woodland and plantation edges, as well as in cultivated areas. It may also be seen from the road, where it perches on exposed branches of old trees and on telephone poles. Numbers have declined in recent years.

BRAHMINY KITE *Haliastur indus* 45cm

This fairly common brown and white kite is mainly seen soaring on thermals, occasionally in small groups. Adult birds are deep reddish-brown with contrasting black primaries; the head and neck are white. It has dull yellow feet and a pale greenish-grey bill. In flight, long broad wings and longish tail distinguish it from the White-bellied Sea Eagle. Mainly a scavenger feeding on carrion, it occasionally takes small mammals, frogs and birds. It has a shrill mewing call. Generally associated with wetlands, coasts, rivers, lakes, mangroves and mudflats. Found throughout the region, but rare in western Java.

WHITE-BELLIED SEA EAGLE *Haliaeetus leucogaster* 68cm

Gerald Cubitt, right

Adult

Juvenile

One of the very large eagles found in the area, this species is distinguished by its magnificent white head, neck and underparts. The rest of the plumage is grey, apart from the black primaries. Immatures show buff where adults are white, the rest of the plumage being dark brown. It flies with slow, powerful wingbeats and occasional glides. It feeds on fish, which it grabs in its talons from the water surface after a swift spectacular dive. Occasionally feeds on refuse. This eagle can often be seen perched upright in trees or on cliffs. It is a fairly common resident, associated with rivers and lakes and especially wooded coasts and rocky shores.

CRESTED SERPENT EAGLE *Spilornis cheela* 54cm

A common eagle, that is often seen over woodland and forests, and often attracts attention by its loud, shrill call, *cwee-chee, chee-chee, chee-cheee*. It is dark-coloured, the upperparts being a greyish chocolate-brown and the underparts similar, but the belly and flanks are spotted with white. It also has a raised patch of white-spotted feathers on the top and back of the head, giving the appearance of a crest. In flight, the broad white tail-band and the band of white on the underwing-coverts are significant. The juveniles appear paler below and on the head. It has yellow feet and a greyish-brown bill. Frequently sits in shade high in a tree, while observing the landscape below.

BESRA *Accipiter virgatus* 33cm

Ray Tipper

Males have uniform dark grey-brown upperparts with a dark greyish-brown head, and a white throat with a black mesial stripe. The breast and belly are rufous-grey, heavily streaked with black and white in the centre of the upper breast; the lower belly is barred white, leading into white undertail-coverts. The tail has a pale tip and shows three wide black bars. Females are generally browner. Sumatran birds (*A. v. vanbemmeli*) are more rufous. The bill is black and the cere grey. This quiet forest bird is confined mainly to mountain and foothill forests, where it preys on birds and reptiles. Resident and fairly widespread throughout the region, but its population is very sparse.

BLACK EAGLE *Ictinaetus malayensis* 70cm

C. Inskipp, BirdLife International

This very dark eagle with its long wings and long tail appears huge in flight. Apart from its yellow feet, yellow cere and the grey tip on its bill, it is all black. There is slight shading on the tail feathers, giving a barred appearance. In flight, a pale area is apparent on the underwing-coverts. Immatures are generally paler, with buff streaking. This eagle feeds on mammals, birds and some carrion, and is known to raid the nests of other birds for both eggs and young. It prefers lowland and the lower hill forests, where it can be seen gliding and circling, often in pairs, over the treetops in hunting forays.

JAVAN HAWK EAGLE *Spizaetus bartelsi* 60cm

Indonesia's national bird, this raptor is easily identified by its very obvious black crest. It has a black moustachial stripe and crown, buff throat and underparts streaked black on neck and brown on belly, and white-feathered legs, while the sides of the head and nape are a rich russet-brown, blending into dark brown upperparts. The tail is brown with black barring. Immatures lack the streaking on the throat and breast, having plain reddish-brown underparts. A rare endemic, found only in hill and mountain forests and open wooded areas in western Java, and down to the coast at Meru Betiri National Park in the east. A last stronghold is in the Gunung Halimun National Park.

BLACK-THIGHED FALCONET *Microhierax fringillarius* 15cm

This minute falcon has black upperparts, but shows white spotting on the tail and secondaries. The crown is black, edged with a white patch above the bill and a white stripe behind the eye, and the sides of the face and ear-coverts are black. The chin and belly are rufous with white margins. Although primarily insectivorous, feeding on dragonflies and grasshoppers, it has also been known to attack larger prey, including small birds. It lives along edges of forests and mangroves and in open country and scrub, and is sometimes seen hunting over paddyfields. Now quite rare in Bali and Java, it has nevertheless managed to survive in reasonable numbers in the wooded lowlands of Sumatra.

WANDERING WHISTLING-DUCK *Dendrocygna arcuata* 45cm

Also called the Whistling Tree-duck, this species makes a series of high-pitched whistling and twittering calls in flight. Its plumage is generally a deep chestnut-brown on the back, tail and breast, but with white undertail-coverts and rump, and a line of black-edged white feathers showing below the folded wing. The head and neck are a paler brown, and there is a dark brown elongated cap running down the back of the neck. The legs are blackish-brown and the bill black. Fairly common throughout the region on wet marshland and on freshwater lakes, where it can be seen diving for food.

Morten Strange

LESSER WHISTLING-DUCK *Dendrocygna javanica* 41cm

Very similar in appearance to the Wandering Whistling-duck, this species is slightly smaller and has no black and white feathers showing below the wing. Its back, crown and underparts are a reddish-brown, but the head and neck are buff. There is often buff streaking along the flanks. The feet are dark grey and the bill black. It calls in flight with a repetitive shrill whistling phrase. It frequently uses trees to roost and rest, and normally nests in tree holes. This is a gregarious duck, usually found in small groups in paddyfields, swamps and mangroves and on lakes. It is resident throughout the region.

WHITE-WINGED DUCK *Cairina scutulata* 75cm

With an estimate of only about 250 individuals worldwide, this large black and white wood duck is seriously endangered. It is a dark-coloured duck with a white head, that of the female being spotted with grey. The back is blackish with green iridescence and the underparts brown. The wing has white lesser-coverts and a greyish-blue speculum. The bill is yellow, tipped black, and the feet dirty yellow-orange. A few of these ducks are located in West Java, and the old logging area at Way Kambas in south-east Sumatra is one of the places where it still hangs on, albeit in small numbers. Here, the dense lowland swamp forest provides conditions to its liking.

COTTON PYGMY-GOOSE *Nettapus coromandelianus* 30cm

Unlike most wildfowl, this species regularly perches in trees and normally nests in tree holes. It is a small black and white duck, the male being predominantly white, with iridescent black plumage on the crown, back, wings and tail and a black neck band. The female's plumage is more subdued, being buffish-white where males are white and brown where males are black; she has a brown eye-stripe and lacks the neck band. In flight, males show a white wing patch. This bird prefers a habitat of marshland and lakes, as well as flooded paddy fields and grassland. It occurs in south Sumatra and West Java.

SUNDA TEAL *Anas gibberifrons* 42cm

This small grey-brown duck is probably the commonest duck of the region, but particularly so in Java and Bali. It is easily distinguished by its unusual bulging forehead. Its sides and back are russet-brown, and it shows a blackish, iridescent blue-green wing speculum. The head and neck are pale buffish-brown, with a dark brown crown. It has grey legs and feet, and the bill is pale grey-blue, turning yellowish towards the tip. It can often be located near water at night by the female's eerie cackling call. Usually in pairs or small groups, often way inland on ponds, rivers and lakes, as well as in mangroves.

PACIFIC BLACK DUCK *Anas superciliosa* 55cm

The name 'black duck' is a misnomer, for this bird's body plumage looks black only in flight when contrasted with the conspicuous white underwing plumage. It is actually dark brown, with the head striped black and white. The speculum is iridescent green and purple, the legs are yellow-brown and the bill is grey. It is a surface-feeding duck which dabbles in shallow water. A resident, confined mainly to the mountain lakes of East Java and Bali, but often found feeding on marshy areas and grassland. A favourite haunt for it in Bali is the sewage farm at Nusa Dua.

BLUE-BREASTED QUAIL *Coturnix chinensis* 14cm

Although the male of this species is distinctive, females are easily confused with the Small and Barred Buttonquails that may be found in the region. The Blue-breasted Quail's yellow feet are the main recognition feature. The upperparts are dark brown with lighter streaks, the female being a little lighter than the male. Males have a distinct black and white bib pattern, with the breast, sides of the head and flanks a rich grey-blue; the belly and undertail-coverts are bright chestnut. The female's underparts are brown, streaked buff, with darker brown barring across the chest; she has a whitish throat patch and buffish eye-stripe. The newly hatched young are tiny, almost like bumble-bees.

CHESTNUT-BELLIED PARTRIDGE *Arborophila javanica* 25cm

Endemic to Java, this partridge occurs in three subspecies. Confined to the west and central parts of the island, each has a different head pattern of reddish-buff with black markings and a black collar. The breast is grey, varying through chestnut to a white vent. The flanks are chestnut, the back and tail grey-brown and barred black, and the wing brown with black barring and small white spots. The feet are red and the bill blackish. This is a bird of montane forest and is often seen in more open spaces, usually in pairs but occasionally in small groups, foraging among the undergrowth.

CRESTED PARTRIDGE *Rollulus rouloul* 25cm

The very distinctive male sports an ostentatious tufted, spiky red crest above a white crown patch. His overall plumage is blackish-blue with a purple sheen, appearing green towards the tail; the wings appear dark red-brown. Females lack the crest, and have green body plumage and a grey head; the wings are more chestnut. They both have red legs and bare red skin around the eye. The bill is black, but males show some red at the base. Congregates in close family groups, foraging through the littered forest floor for insects and fallen fruit. Found in the lowland and hill forests of Sumatra, but not in Java or Bali, the species also occurs in Borneo and the Malay Peninsula.

CRESTED FIREBACK *Lophura ignita* 55cm

It is extremely unfortunate that this beautiful pheasant, once common in the forests of Sumatra, Borneo and the Malay Peninsula, now occurs only sporadically. Destruction of the habitat and hunting have reduced its numbers to isolated populations in the more remote and secluded forests. Sumatran birds have a very dark blue, almost black, body with a tufty black crest and a red patch on the rump and lower back; the belly and flanks are faintly streaked with white. The extended and arched central tail feathers are white. Females are brown, with white-striped underparts. These are ground-living birds which forage like chickens for fallen fruit, seeds and insects.

Male (above); female (below)

RED JUNGLEFOWL *Gallus gallus* male 75cm, female 46cm

Male (above); female (below)

Chew Yen Fook

Two subspecies of this wild ancestor of the domesticated fowl exist in the region: *G. g. spadiceus* in north Sumatra and *G. g. bankiva* in south Sumatra, Java and Bali. Males are recognized by their serrated red comb, face and wattles, their long bronze hackles (longer in the north Sumatran race) and their long, dark green arched tail feathers; they have a chestnut to golden mantle, and blackish-green breast and primary coverts. The females are much duller, being various shades of brown, with black streaking on the neck. The bill is buff and the legs slate-grey. Found in the more open, scrubby areas of forest edges and clearings.

GREEN JUNGLEFOWL *Gallus varius* male 60cm, female 42cm

Endemic to Java, Bali and Nusa Tenggara, this large blackish-green fowl is very similar to the Red Junglefowl but with an unserrated purplish comb. Males have red wattles and bare skin around the eye, a glossy green mantle and nape and iridescent green hackles, with uppertail and wing-coverts orange and yellow; the rest of the plumage is black. Females have brown upperparts sparsely mottled with buff, and buffish underparts mottled with black. This bird prefers the more open grassy areas and often associates with grazing animals, which disturb insects on which it can feed.

GREAT ARGUS *Argusianus argus* male 120cm, female 60cm

The Great Argus is one of the most spectacular of the region's pheasants. The males have extraordinarily elongated secondary and tail feathers, used in their mating display to attract females. With tail raised high and the wings spread wide, the beautiful patterning of green 'eye' marks (ocelli) is shown off to perfection. The plumage is otherwise mainly chestnut-brown, broken and patterned by spots and flecks of buff and black. The female is generally darker, has shorter wings and tail, and lacks the male's 'eye' spots. Both sexes have a short dark brown crest and bare blue skin on the head and neck.

GREEN PEAFOWL *Pavo muticus* male 210cm, female 120cm

This spectacular pheasant can be found roaming forest edges and savanna woodland by day, roosting in bare trees at night. Predominantly iridescent green, the males of this species have conspicuous ocellated tail feathers and a vertical crest on the head. Females are similar, but with shorter tail and paler underparts. During the breeding season, their mewing and loud pairing calls are very obvious and the male's striking display with tail feathers fanned is unmistakable. Once a familiar sight in open woodland, pastures and plantations, this beautiful bird has been virtually exterminated by hunting. Within the region, Java's Ujung Kulon and Baluran National Parks are the last remaining strongholds for this species, but a few local relict populations exist elsewhere.

SLATY-BREASTED RAIL *Gallirallus striatus* 25cm

Not only is this a very secretive and mainly solitary bird, it is also partly nocturnal and therefore seldom seen. Distinguished from other rails in the region by its chestnut crown, it has a grey upper breast and white chin. The upperparts are brown with fine black and white barring; lower-breast and underparts are white, barred with black. The bill has a pink base and a grey tip, while the legs and feet are grey. It is a common bird of low wetland habitat such as mangroves and marshy areas, as well as paddyfields and even drier areas of dense alang-alang grassland.

Chew Yen Fook

RUDDY-BREASTED CRAKE *Porzana fusca* 21cm

This crake is easily confused with three others that frequent the region: the Red-legged Crake (*Rallina fasciata*), the Slaty-legged Crake (*Rallina eurizonoides*) and the Band-bellied Crake (*Porzana paykullii*). On close observation, it is the only one with both head and breast a bright chestnut, with a white chin, and blackish-brown underparts finely barred white. Upperparts are plain russet-brown. Like the other crakes, it is a shy bird and partially nocturnal, preferring the seclusion of paddyfields and reedy habitat and often bushland adjacent to water. A relatively common resident found throughout the region's lowlands.

WHITE-BROWED CRAKE *Porzana cinerea* 27cm

This little crake can be quite noisy by day and night during the breeding season. It is easily distinguished from all other small crakes in the region by its conspicuous white head markings above and below the black eye-stripe. The tail and wings are brown, the back and crown are dark greyish-brown, and the belly, neck and throat are white, becoming greyer at the sides. It is ideally adapted to living in marshland, rice paddies and flooded vegetation, where its proportionately large long-toed feet allow it to move around on top of floating vegetation and pick out aquatic invertebrates. It is, however, very timid and will run for cover at the slightest disturbance.

Chew Yen Fook

WHITE-BREASTED WATERHEN *Amaurornis phoenicurus* 33cm

Chew Yen Fook

The most conspicuous and easily identified rail found in the region. Its white face, foreneck and breast, along with rufous flanks and lower belly, are well demarcated from the dark grey-green upperparts. The bill is greyish-green with a red base, and the legs and feet yellow. It is very vocal at dawn and dusk, making a cacophony of weird squawks, grunts and croaking sounds. It is a wandering feeder, frequently out in the open but never far from dense cover. Very agile, it runs over floating vegetation and clambers through bushes, searching for insects. More often seen alone, but frequently in twos and threes.

COMMON MOORHEN *Gallinula chloropus* 31cm

This very adaptable aquatic bird is almost entirely dark grey to black, with a broken white stripe along its flanks. It also displays white patches under the tail, particularly when dashing for cover with its tail raised. It has a characteristic red frontal shield and bill with a yellow tip. The legs are greenish-yellow. Not only is it a good surface swimmer, it will also dive and even run over the water's surface to avoid predators. It feeds mainly on aquatic insects. It frequents freshwater pools, lakes and rivers, and is naturally 'at home' in the rice paddies of the region. Found on almost every continent worldwide, and is also a common resident of Sumatra, Java and Bali.

PURPLE SWAMPHEN *Porphyrio porphyrio* 48cm

A rather clumsy-looking bird, big and brightly coloured, its large size, and blue-black plumage, offset by a substantial bright red bill and frontal shield making it easily recognized. Its undertail-coverts are white and it habitually flicks its tail. Its long, ungainly pink legs with huge spreading claws are ideal for both trampling through and over aquatic vegetation. Although not uncommon, it is a rather quiet and secretive bird, that keeps to the edges of undisturbed reedbeds, marshland and occasionally paddyfields. It is found throughout the region.

GREY PLOVER *Pluvialis squatarola* 28cm

In winter plumage, this rather attractive wader has upperparts mottled grey and brown and underparts pale grey to white, with a black bill and dark grey legs. In flight, it shows a pale wingbar and rump and, below, a white underwing with a large black patch near the body. These features easily separate it from the Pacific Golden Plover, which is also slightly smaller and is shorter-billed. The Grey Plover breeds in the Arctic and is a winter visitor to the region's coastal areas. Found in small groups on tidal mudflats, where it feeds on marine invertebrates and often roosts on isolated shingle spits.

Tilford/Cooke, T C Nature

PACIFIC GOLDEN PLOVER *Pluvialis fulva* 25cm

Chew Yen Fook

A typical plover in shape, with a robust body, thin neck and largish head with a short strong bill. When in the region it is usually in non-breeding plumage, having lost the distinctive black face and underparts it had on its breeding grounds. It can be identified by its speckled gold, brown and buff upperparts and buff breast, face and supercilium. The bill is black and the legs grey. This species breeds in north Siberia, north-east Asia and Alaska and is a winter migrant to South-east Asia and Australia, stopping off around the coasts of the Sundas. It is common on mudflats and open grassland near the coast, where it often congregates in flocks.

MALAYSIAN PLOVER *Charadrius peronii* 15cm

Morten Strange

One of the less timid waders, this small bird is more likely to be seen alone or in small groups, and it is unusual for it to flock with other waders. Like many of the small plovers, it is black, brown and white and has a short black bill. It is significantly smaller than the Lesser Sand Plover (*Charadrius mongolus*) and Greater Sand Plover and it also has a narrow collar, which is black on males and brown on females. Unlike on the Kentish Plover, its narrow black eye-stripe is separate from the dark brown ear patches. It is quite common on sandy beaches, where it forages along the shoreline.

GREATER SAND PLOVER *Charadrius leschenaultii* 23cm

B. R. Hughes, Windrush Photos

This brown, white and grey plover joins with flocks with other waders, especially the Lesser Sand Plover (*Charadrius mongolus*), on mudflats and sandy estuarine beaches, typically probing the ground for invertebrates. It is most easily separated from the smaller but otherwise similar Lesser Sand Plover by its more substantial bill; apart from the latter, all the other wintering plovers likely to be encountered in the area show a breast-band or a collar. Early arrivals, however, may show remnants of the breeding plumage and have a russet breast bar. This common winter migrant ranges from Africa and Asia through to Australia and New Zealand.

WHIMBREL *Numenius phaeopus* 43cm

Tilford/Cooke, TC Nature

A large, long-legged wader with a fairly long neck and long decurved bill. Its plumage is heavily mottled brown, with a prominent black crown-stripe above a buff eyebrow and dark eye-stripe. In flight, it shows a white underwing, rump and lower back. Normally quiet, it will often give a trilling call when alarmed. A common visitor to the region's coastline, where it congregates with flocks of other waders on tidal mudflats and estuaries. It also occurs on coastal marshes and rough grassland. Non-breeding birds are often present throughout the year.

BAR-TAILED GODWIT *Limosa lapponica* 37cm

Tilford/Cooke, TC Nature

Both this species and the similar Black-tailed Godwit (*Limosa limosa*) can be present at the same time. They are largish, long-legged waders with a fairly straight and long bill. Both have greyish-brown upperparts, more heavily mottled on Bar-tailed, which also has an obviously barred tail and slightly heavier grey streaking on the whitish breast. It is generally found in areas frequented by other waders, especially tidal mudflats, estuaries, open expanses of sandy beach and saltmarsh. This fairly common winter visitor to Sumatra is not nearly so common in Java and Bali. Those seen in the region breed in north-east Asia, and are usually in their non-breeding plumage.

WOOD SANDPIPER *Tringa glareola* 20cm

Chew Yen Fook

This is quite a noisy species, frequently uttering its *chee-chee-chee* call. Its greyish-brown upperparts have contrasting black and white speckling, and its underparts are white, with a greyish breast; the forehead is buff, leading into a narrow eyebrow. It has long yellowish-green legs. In flight, the white rump and underwing are obvious; there is no wingbar. Breeds in north Europe and north Asia, wintering as far south as Australia. A common visitor throughout the region, seen in small groups, in wet muddy areas, not only on coastal mudflats and mangroves but also in marshland and paddyfields far inland.

COMMON SANDPIPER *Actitis hypoleucos* 20cm

Tilford/Cooke, T C Nature

This is a very 'busy' feeder with a constant tail-bobbing action, and when disturbed it often flies a short distance and continues, seemingly unperturbed. It has brown upperparts and white underparts, with a brown patch extending down from the neck to the breast side. In flight, its white and brown barred outer tail feathers are evident, as is the white wingbar. It is most likely to be found alone, feeding on coastal mudflats and along muddy river banks, but occasionally it visits surrounding marshland and paddyfields upstream. A very common migrant from Eurasia, seen throughout the year.

SANDERLING *Calidris alba* 20cm

Tilford/Cooke, T C Nature

This highly active wader can often be seen following the retreating waves as it feeds on marine invertebrates thrown up on to the beach. It usually occurs in small groups. In non-breeding plumage, it has pale grey upperparts and white underparts, with black bill and feet. A black shoulder patch helps to distinguish it from other waders. It shows black primaries and a prominent white wingbar in flight. The Sanderling breeds in the Arctic and occurs as a non-breeding migrant throughout the region, but is not common. Found almost entirely on sandy beaches.

LONG-TOED STINT *Calidris subminuta* 14cm

Ray Tipper

This little wader has a relatively long neck and legs compared with others of its size. In non-breeding plumage, it has heavily streaked brownish-grey upperparts, with a grey-brown crown and bold white eyebrow. The underparts are white, the breast streaked brownish-grey. In flight, the centre of the rump and tail appear black and the outer tail buff. A regular, at times numerous, passage migrant from Siberia, passing through the Greater Sundas on its way to New Guinea or Australia; juveniles occasionally stay through the summer. It usually occurs on coastal mudflats and estuaries, but frequently inland on muddy marshland and rice paddies.

BLACK-NAPED TERN *Sterna sumatrana* 30cm

Kevin Carlson, &Windrush Photos

Apart from the conspicuous black band through the eye and around the nape and its very pale grey upperparts, this small tern appears very white. In flight, its very long forked tail is obvious. The legs are black, and the black bill is tipped with yellow. Juveniles, however, have grey-brown mottled plumage over the upperparts and on the crown, with black mottling on the nape; the bill-tip is also brownish and the tail unforked. This quite common resident breeds on offshore rocky islands, and is generally seen in the company of other terns on sandy shorelines.

BRIDLED TERN *Onychoprion anaethetus* 37cm

Arnoud B. van den Berg, Windrush Photos

Although resident in the area, Bridled Terns spend most of their time out at sea, being driven ashore by bad weather or by the necessity to breed and moult. They feed on small fish or floating invertebrates, taken from the surface of the sea. This is a medium-sized tern with dark grey-brown back, tail and wings, white underparts and a long forked tail. The crown, nape and eye-stripe are black, leaving a thin white forehead leading into a short white supercilium. The leading edge of the wings is white, as are the outer tail feathers. Seen ashore more during the summer months, but then only singly or in small groups.

GREAT CRESTED TERN *Thalasseus bergii* 45cm

This tern appears very large, and has a grey back and rump and white underparts, neck and patch above the bill. In breeding plumage, it has a black cap extending backwards to form a slight crest. During the summer, the black crown becomes heavily mottled with white, and gradually turns to white-mottled grey as winter approaches. It has a yellow bill and black feet, the bill colour distinguishing it from the very similar but slightly smaller Lesser Crested Tern. Juveniles are heavily marked with brown on the upperparts and have a dark grey tail and outer primaries. A very common tern of inland waters.

LESSER CRESTED TERN *Thalasseus bengalensis* 40cm

Steve Young

This species looks very similar to the Great Crested Tern, but is slightly smaller and has a distinctive orange bill. In breeding plumage, the black cap extends forward to meet the bill. After breeding, the forehead and forecrown become white, leaving a black crest. Juveniles appear more like non-breeding adults, but with greyish wings and mottled brown upperparts. Often associates with other terns, sometimes in large flocks. Occurs in coastal waters and often far out to sea, but is more often observed along sandy and muddy shorelines. A common winter visitor to Java and Bali, but less so to Sumatra.

THICK-BILLED GREEN PIGEON *Treron curvirostra* 27cm

Chew Yen Fook

This is a fairly robust pigeon with red feet and a yellow iris surrounded by a pale bluish-green eye-ring. Males have a maroon mantle and back, whereas females are green. The underparts are bright green, the flanks darker green and barred white. The neck is green, becoming greyish on the ear-coverts, with the forehead and crown grey. The two subspecies in Sumatra (*T. c. curvirostris* and *T. c. harterti*) have a substantial pale green bill with a red base, while that of *T. c. hypothapsina* from the islands of south-west Java has an olive base. A fairly common and noisy bird of lowland forest, especially among fruiting trees.

PINK-NECKED GREEN PIGEON *Treron vernans* 27cm

Tilford/Cooke, TC Nature

This small green pigeon is distinguished from similar pigeons by its grey tail with black band and pale grey tip. The male has a blue-grey head merging through pink to orange on the lower breast; the abdomen is green and yellow, and the back and wings green. The female is green, lacking the male's brighter colours, but identified by tail pattern and by its company with the distinctive male. Its calls are a 'cooing' whistle and, when feeding, a series of crow-like rasps. Common in lowland and coastal forest as well as in more open countryside, becoming ever more conspicuous when trees are fruiting. Occurs from southern Burma through to the Lesser Sundas, and normally resident.

BLACK-NAPED FRUIT DOVE *Ptilinopus melanospilus* 27cm

Tilford/Cooke, TC Nature

This species usually occurs in pairs but being very timid it is seldom seen except when it congregates in flocks at fruiting trees. It is, however, usually detected by its resonant *ow-wook-wook... ow-wook-wook* call. It has predominantly green upperparts, tail and lower breast with a yellow vent and red undertail coverts. The male has a pale grey head with a conspicuous black nape and yellow throat. Females have all green heads. Both have red feet and greenish-yellow bill. It is a locally common fruit dove of the lowland and hill forests of Java and Bali not normally found on Sumatra.

JAMBU FRUIT DOVE *Ptilinopus jambu* 28cm

Predominantly bright green upperparts, white underparts and red face separate this from other fruit doves. The male's face is crimson, and he has a black throat patch, a bright pink tinge to the upper breast and chestnut undertail-coverts. The female has a dull purple face, her undertail-coverts are orange-brown and the upper breast bright green. The bill is pale orange and the legs crimson. A common lowland bird of Sumatra, it is rarely found beyond West Java. It prefers more open wooded areas with an abundance of fruiting trees. It occurs mainly around the coast and on small offshore islands.

Chew Yen Fook

49

GREEN IMPERIAL PIGEON *Ducula aenea* 42cm

Chew Yen Fook

A large green and grey pigeon which prefers to live high in the treetops, where it feeds on figs and nutmeg and other small fruits, normally swallowing them whole. It is the commonest pigeon of the region's lowland rainforest, but is becoming increasingly more difficult to locate as its habitat is lost. Head, neck and underparts are soft pinkish-grey, with chestnut undertail-coverts. The upperparts are green with a bronze iridescence. It has a dull red base to the blue-grey bill, and dark red feet. This species inhabits coastal mangroves and, especially, riverine forest. It occurs from India through the Sundas to New Guinea, and is resident.

PIED IMPERIAL PIGEON *Ducula bicolor* 40cm

Chew Yen Fook

This is a striking bird, mainly creamy to ivory-white, with black flight feathers and tip of tail; the bill and feet are blue-grey. Immatures are greyer. Found more often in mangrove and coastal forests, it sometimes congregates in substantial breeding colonies where fruiting figs are readily available. Hunting has decimated the population of this beautiful pigeon, particularly in Java and Bali. Fortunately, it still remains quite common on the offshore islands of Sumatra and elsewhere in its range, which extends from the south of Burma to New Guinea. It is a resident species, but often moves long distances between islands.

RUDDY CUCKOO DOVE *Macropygia emiliana* 30cm

This species is sometimes treated as an Indonesian form of the Brown Cuckoo Dove (*M. amboinensis*). It is much smaller than the Barred Cuckoo Dove (*M. unchall*) and more of a chestnut-brown. The male's upperparts are uniform brown, with a light purplish iridescence on the neck and upper breast. The underparts are slightly paler chestnut-brown. Females are similar, but have black barring on the back and wing-coverts and slight barring on the upper breast. It prefers primary-forest edges and clearings, where it spends much of its time in search of food. Common in the hill forests of Java and Bali, with endemic forms occurring on the south Sumatran islands, although there are few records from Sumatra itself.

LITTLE CUCKOO DOVE *Macropygia ruficeps* 29cm

Morten Strange

Smaller than the other brown cuckoo doves, the Little Cuckoo Dove is separated by its buff breast and by the dark brown subterminal bar on its outer tail feathers. The upperparts also are lightly barred black and the upper breast is often heavily barred, especially on the female. Males have green and purple iridescence on the nape. They have red legs and a brown-tipped black bill. This species congregates, often in large flocks, along forest edges, feeding in grassland and ripening rice fields. It is a common bird in the hills and montane forests throughout the region.

SPOTTED DOVE *Streptopelia chinensis* 30cm

This very familiar dove has a generally pinkish-brown plumage, with slightly darker flight feathers and dark-mottled back. On each side of the neck it has an obvious white-spotted black patch. The outer tail feathers are broadly tipped white, this being very obvious in flight. It has red feet and a black bill. It is frequently found close to human habitation, from city gardens to rural cultivation, open country and forest edges in lowland areas, where it forages on the ground. Very common throughout the Sundas, and widely kept as a cagebird and bred in aviaries throughout the world.

PEACEFUL DOVE *Geopelia striata* 20cm

A very common cagebird throughout Indonesia, but particularly in Java, where it holds a special significance in Javanese philosophy. Also known as the Zebra Dove for its finely barred pale brown and black plumage from neck to tail. The back of the head is pale brown, blending into a buff face, crown, neck and chin. The long tapered tail is brown and black with white-tipped outer feathers; very evident on take-off. This is a ground feeder, occurring around gardens, cultivation and open grassland and scrub. Although common in lowland south Sumatra, excessive trapping in Bali and Java has dramatically reduced its numbers.

EMERALD DOVE *Chalcophaps indica* 25cm

Chew Yen Fook

This attractive ground dove has iridescent emerald-green wing-coverts and mantle, with a dark grey lower back crossed by two conspicuous white bars. The underparts, neck and sides of the head are dark pink, and the short tail is black. Males have a white forehead, with the crown and nape tinged grey. Females are duller. The species occurs singly or in pairs, preferring the seclusion of thick forest or forest edge. Throughout its range from Australia to India it is quite common, but it is becoming rare in Java and Bali, probably as a result of habitat destruction and trapping for the pet trade.

NICOBAR PIGEON *Caloenas nicobarica* 40cm

One of the more unusual pigeons, recognized by its long, iridescent purple, mane-like hackles. Apart from the short, stubby white tail, the plumage is dark green, with an iridescent sheen of gold, purple and bronze on the upperparts. The bill is black, with a conspicuous knob-shaped cere. The legs are purplish-red and the claws yellow. This is a wary bird and almost silent, making an occasional deep croaking call. Although it is a secretive ground feeder, it rests low down in trees for the greater part of the day, only becoming active at dusk. It is fairly rare, occurring mainly on the offshore islands of Java and Sumatra.

RAINBOW LORIKEET *Trichoglossus haematodus* 24cm

This is a brilliant multicoloured parrot with a very dark purplish-black head. Apart from the yellow collar, the upperparts are green. It has a red breast and underwing-coverts, and a yellow patch under the shoulder. The belly is very dark purple, and the thighs show alternate green and yellow narrow bands. The feet are grey and the bill red. It is normally seen in small, noisy, foraging parties flying low over the forest canopy. A resident species in Java, Bali and the Lesser Sundas through to Australia. Once common in Bali, it is now quite rare, having been exploited for the pet trade.

Tilford/Cooke, TC Nature

YELLOW-CRESTED Cockatoo *Cacatua sulphurea* 33cm

A fairly large, white, parrot-like bird with a lovely crest of stiff yellow feathers and yellow cheek patches. It has a black bill and dark grey feet. Its flight appears very laboured, with alternate glides and fast flapping wingbeats. It can often be seen in small groups, but usually lives in pairs, and is very noisy, with screeches, squawks and whistles in its repertoire; it has a habit of raising its crest as it calls. Also known as the Lesser Sulphur-crested Cockatoo, this bird is endemic to Sulawesi and the Lesser Sundas, with a population on Nusa Penida, Bali, and another on Mesalembu Besar in the Java Sea. Has been widely exploited by the pet trade.

Tilford/Cooke, TC Nature

YELLOW-THROATED HANGING PARROT
Loriculus pusillus 12cm

These very small and well camouflaged parrots are more often seen in the treetops, clambering around and hanging, often upside-down, to feed on flowers and small fruits. They fly around in small flocks, often making high-pitched shrieking calls. They are mainly leaf-green, with brighter yellow-green underparts, and are particularly difficult to observe as their colouring blends in so well with the foliage. Males have a small yellow patch on the throat. The feet and bill are orange. A fairly common endemic of Java and Bali, inhabiting lowland rainforest.

Tilford/Cooke, TC Nature

SUNDA CUCKOO *Cuculus lepidus* 26cm

This very secretive species spends much of its time high in the forest canopy, and is very quiet outside the breeding season. In February and March, it can be heard giving its three-syllable *hoop, hoop-poop* call. This quite small cuckoo has a grey breast and upperparts and a dark grey tail; the underparts are buff, barred black. It has a yellow eye-ring, grey bill and dirty orange-yellow feet. There is an hepatic form of the female which has black-barred rufous upperparts and white-barred underparts, very similar to the juveniles. Formerly regarded as a race of Oriental Cuckoo (*C. saturatus*), which also occurs during the winter.

Tilford/Cooke, TC Nature

DRONGO CUCKOO *Surniculus lugubris* 24cm

This predominantly black cuckoo has a forked tail like a drongo. Its plumage is shiny black, apart from white bands on the undertail-coverts and all-white thighs. The bill is black and the legs are grey. The plumage of the juvenile is flecked with white and the white tail-bands are wider. Although shy, this species has a loud, clear call consisting of an ascending scale of up to seven notes and less often a series of warbled notes. It occurs in the lowlands throughout the region, both as a resident and as a migrant, but its timid nature confines it mainly to forests and forest edges.

ASIAN KOEL *Eudynamys scolopaceus* 42cm

Male (above); female (below)

This bird's loud call, a shrill repeated *koel, koel, koel* ..., is the origin of its name. Its other call, a loud and accelerating *kow-wow, kow-wow, kow-wow* ..., can be heard by day and night, but you will be lucky if this very timid bird reveals itself. Like others of the cuckoo family, it is a brood-parasite, searching out nests of orioles, crows and drongos in which to deposit its eggs. The adult male is glossy black, with a long tail, buff bill and red eye-ring. The female is grey to golden-brown, with heavy white speckling and barring.

RED-BILLED MALKOHA *Phaenicophaeus javanicus* 46cm

Separated from other malkohas of the region by its large size and strong red bill. The upperparts are mid-grey but tinged with a pale bluish-green iridescence. The underparts are chestnut with a grey band across the upper breast and the grey tail feathers are tipped with white. The eye is surrounded by a bare patch of blue skin and the legs are grey. Found in the lowland and lower hills of Java and Sumatra, it prefers forest edges and secondary scrub. It usually moves about in pairs or small family groups often foraging amongst alang-alang grassland. It has a 'whining' call.

Tilford/Cooke, TC Nature

CHESTNUT-BREASTED MALKOHA *Phaenicophaeus curvirostris* 50cm

The tail of this malkoha is over half the bird's total length, appearing dull green above with a rufous tip and rufous on the underside. The upperparts are also dull green, with the crown and nape grey. There is a bare red patch of facial skin surrounding the eye. The Sumatran race has grey throat and cheeks and a black patch across the lower belly. There is also plumage variation in other island races. This locally common lowland bird is usually found among forest thickets and can often be observed perched stationary in the tops of small trees.

Morten Strange

GREATER COUCAL *Centropus sinensis* 52cm

Chew Yen Fook

Although often on the ground, this large, black, crow-like bird with chestnut-brown wings is more likely to be seen lumbering about in small trees and bushes, where it can hide in thick vegetation. The bill and feet are black. Its very eerie, far-reaching, low-pitched call starts with a series of *poop, poop, poop* sounds, increasing in tempo and then slowing again, but occasionally a series of just four deep monotonous *poop, poop, poop, poop* notes is given. It is heard more often in the morning. A resident throughout the Sundas, preferring forest edges, mangroves, grassy banks and secondary scrub.

LESSER COUCAL *Centropus bengalensis* 40cm

Morten Strange

Very similar in appearance to the Greater Coucal, this species is smaller and has duller plumage, and is also more common throughout the region. Its flight is weak, consisting only of short, laboured, flapping forays low over the vegetation in search of insects. It hops when on the ground, and clambers around when in trees and bushes. Like other coucals, the plumage is black except for the chestnut back and wings; the bill and feet are also black. Found especially in the lowlands, where it frequents open grassy areas, particularly the denser alang-alang grassland and marshland.

INDIAN SCOPS OWL *Otus bakkamoena* 20cm

This common owl can often be seen hunting for insects and small rodents from trees and prominent perches in built-up areas of towns. The regular soft *whoop* call of the male and the *weeoo* and *plop* calls from the female are often heard alternately as a continuous duet. Identification is not simple, as the species occurs in several colour forms. It can be dull reddish-brown to grey-brown, with the upperparts mottled and blotched with buffs and black; the underparts are paler and streaked with black. It has a pale sandy 'neck collar'. Bill and feet are yellow. Found almost everywhere below 1500m and possibly the commonest owl encountered.

Tilford/Cooke, TC Nature

BARRED EAGLE OWL *Bubo sumatranus* 45cm

One of the larger owls to be found in the region, this magnificent bird is unfortunately rarely seen, confining itself to the lowland forests and emerging from its daytime roosting place only at dusk. Recognized by its large size, its prominent horizontal ear-tufts, and its pale grey underparts heavily barred with black. The upperparts are dark brown, flecked and barred with buff. The eyes are brown and the feet and bill yellow. It has many unusual calls, which are often thought to be attributable to mysterious demons by the local people. Its familiar call is a loud *whooa-hoo, whooa-hoo*, followed by a deep groan.

Chew Yen Fook

59

BUFFY FISH OWL *Ketupa ketupu* 45cm

This large owl has an amusing appearance with its bright yellow eyes and its conspicuous horizontal ear-tufts. Its upperparts are brown, streaked with black and buff, and the underparts are bright rufous-buff with narrow black streaks. It has whitish eyebrows, a grey bill and yellow feet. It is a mainly nocturnal species, preferring to remain hidden among trees during the day. At night, it ventures outside the forest into parks, paddyfields and often alongside roads. It is never far from water, on which it is dependent for its fishing for frogs, fish and other aquatic life.

SPOTTED WOOD OWL *Strix seloputo* 47cm

Being nocturnal, this bird is seldom seen, but its presence is disclosed by its low growling and resounding *hoo-hoo-hoo* calls. It is a large owl, similar in shape to the Barred Eagle Owl, but without the ear-tufts. Its facial discs are light rufous and the underparts are whitish, tinged rufous and lightly barred with dark brown. A white throat is characteristic. The upperparts are a rich chocolate-brown, marked with black-edged white spots. The bill is greenish-grey and the feet grey. Found in the lowland forests of Java, it may also be attracted to woodland close to villages and sometimes into towns.

LARGE-TAILED NIGHTJAR *Caprimulgus macrurus* 30cm

Mike McKavett, Windrush Photos

This quite large grey-brown nightjar spends the daytime in shady areas on the ground, concealed by its camouflaged plumage. Its activity commences at dusk when, for half an hour, either perched in a tree or on the ground, it utters its slow and repeated *choink-choink-choink* call. In flight, the broad white tips of the two outer pairs of tail feathers and the distinctive white patch covering the centres of the four outer primary wing feathers aid identification. Widespread from India to Australia, this bird is locally common throughout the region, where it prefers lowland wooded country, mangroves and forest edges.

SAVANNA NIGHTJAR *Caprimulgus affinis* 22cm

One of the smallest nightjars in the region, this species has a monotonous repeated call, *jweep*, which is heard around dawn and dusk, often from birds in flight. Its fairly uniform but well-camouflaged brown plumage has a white patch on each side of the neck; the male has white outer tail feathers. It is common in dry open areas around the coastal lowlands, but is found also, somewhat surprisingly, in big cities, where it exploits the hordes of flying insects attracted by the lights and uses the large flat rooftops of buildings on which to rest. In open country it roosts on the ground by day, and is seldom seen perched elsewhere.

61

GLOSSY SWIFTLET *Collocalia esculenta* 9cm

Chew Yen Fook

This is the smallest and commonest swiftlet in the region, identified by its glossy black upperparts and whitish belly. It also goes by the alternative name of White-bellied Swiftlet. The chin is pale grey, and in flight a small notch is apparent in the tail. Both bill and feet are black. It nests in the entrances to caves, but it does not use echolocation and the nest is inedible. Like all swifts it is insectivorous, catching aerial insects as it flies over forests and agricultural land, from the coastline to the highest peaks. It often flies over water, dipping down to drink. Widespread throughout the region.

HOUSE SWIFT *Apus nipalensis* 15cm

This medium-sized swift, recently split from Little Swift (*A. affinis*), is usually seen in large groups hunting insects over open countryside. It can be recognized by its white rump and throat and slightly notched tail, unlike the similar but larger Fork-tailed Swift (*Apus pacificus*), which has a distinct fork to the tail. Otherwise it has blackish plumage, black bill and brown feet. It prefers coastal habitats up to the lower hills and is frequently found in towns and villages, where it nests under the eaves of houses; it also uses overhanging cliff faces and cave entrances as nesting sites. It is locally common over much of the region.

Morten Strange

COMMON KINGFISHER *Alcedo atthis* 15cm

Tilford/Cooke, TC Nature

This is the common kingfisher seen from Eurasia all the way to New Guinea. Upperparts are a shiny bright blue-green, with paler blue uppertail-coverts and a small patch in the centre of the back. The underparts are rufous, as are the ear-coverts. There is a white stripe on the side of the neck. Feet are red and the bill black. In flight, it has a drawn-out high-pitched *tseeep, tseeep* call. Frequents lowland open country with freshwater mangroves, rivers and lakes. Uses an overhanging branch or rock beside water as a perch from which to dive for fish. A common visitor to Sumatra, but less so to Java and Bali.

BLUE-EARED KINGFISHER *Alcedo meninting* 15cm

This is the woodland counterpart of the Common Kingfisher, and is very similar in its behaviour, moving between waterside perches in rapid flight. It dives into water to catch fish, which it kills by bashing it on the perch before swallowing it whole head-first. The Blue-eared has a more contrasty appearance than the Common Kingfisher, its upperparts being a darker but shinier blue and its underparts a bright orange-red. The ear-coverts are blue. It frequents forest rivers, streams and lakes, and occasionally estuarine habitat, throughout the lowlands of Sumatra, Java and Bali.

63

SMALL BLUE KINGFISHER *Alcedo coerulescens* 14cm

Morten Strange

Also known as the Small Blue Kingfisher, this tiny species is easily identified. The upperparts are a brilliant pale greenish-blue, with the crown, wing-coverts and a neck patch barred with dark blue; it also has a pale greenish-blue band across its upper breast. The throat, lores and belly are white. In typical kingfisher fashion, it perches on prominent places above water, waiting to dive in on its prey. It is quite common on estuaries and in the coastal mangroves and marshland of Bali and Java, reaching west into southern Sumatra. A good place to see many of the kingfishers is around the numerous coastal fish-farms.

BLACK-BACKED KINGFISHER *Ceyx erithaca* 14cm

Morten Strange

This brilliantly coloured little kingfisher has adapted to feeding on insect prey and lives in forests often some distance from water. It has a rapid low flight, during which it often emits a very high-pitched whistle. It hunts from a perch, and often snatches spiders from their webs while in flight. It has distinctive brilliant yellow underparts and bluish-black back and wing-coverts. The crown and tail are red, as are the bill and feet. There are usually small patches of blue in front of the eye and behind the ear-coverts. Widespread in suitable habitat throughout the region but not common.

STORK-BILLED KINGFISHER *Pelargopsis capensis* 35cm

This kingfisher, too, being the largest encountered in the region and having an enormous red bill, is easy to identify. The back, wings and tail are blue-green and the underparts pinkish-orange. The head is brown, becoming lighter around the neck. It has short red legs. It inhabits coastal mangroves and lowland rivers and marshland, as well as agricultural land, paddyfields and woodland adjacent to rivers. It feeds on fish, crabs and amphibians. Once recorded throughout the area, it is now an uncommon resident, and possibly extinct in Bali. Farming practices, pesticides and disturbance are thought to be the main reasons for its decline.

WHITE-THROATED KINGFISHER *Halcyon smyrnensis* 27cm

Ray Tipper

The white throat and 'bib' contrasting with the chocolate-brown head and underparts are distinguishing features for this large kingfisher. The upperparts, tail and upper surfaces of the wings are iridescent mid-blue, and there is a patch of brown on the upperwing-coverts. A noisy loud screaming call, *kee-kee-kee-kee*, emitted while perched or in flight, announces the bird's presence as it hunts over rivers and ponds and even along the coast. Found mainly in Sumatra, with a few records in West Java, this kingfisher inhabits mainly lowland open areas near to water.

JAVAN KINGFISHER *Halcyon cyanoventris* 25cm

This dark-looking kingfisher has adapted well to a diet of insects and terrestrial prey, and is not reliant on fishing for its survival. It can often be seen on an isolated perch in open grassland, waiting for insects to come into range, although it remains near water where prey items are more abundant. It is dark brown on the head, with lighter brown collar and upper breast. The belly and hindcollar are purplish-blue, and the back and wing-coverts dark purple. Flight feathers and tail are bright blue. The white wing patches are particularly conspicuous in flight. Both bill and feet are red. Endemic to lowlands in Java and Bali.

COLLARED KINGFISHER *Todiramphus chloris* 24cm

A very familiar kingfisher, this species is quite evident from its continual calling throughout the day. Its harsh, noisy *chue-chue-chue-chue-chue* carries over long distances and announces its presence almost everywhere except in the mountains. At first sight it is bright blue and white, but at close range the blue of the crown, back, wings and tail has a superb green iridescence. It has a beautifully clean white collar and underparts. The lores are white, and a black stripe passes through the eye and around the back of the head. It prefers open country around the coast and wherever there is water in which to fish, and must be the commonest kingfisher of the region.

CHESTNUT-HEADED BEE-EATER *Merops leschenaulti* 20cm

Like all bee-eaters, this species congregates in parties which wander around searching for insect prey, which it catches in flight. These birds habitually choose exposed perches from which to survey and make their sweeping excursions to catch insects, bringing them back to the perch to break up and eat. It has a light chestnut crown, nape and mantle, a yellow throat and below this a chestnut and black band. In flight, the pale blue rump contrasts with the dark green back and tail, and the underwing appears orange. The tail is square-ended and lacks the streamers of similar species. Quite common in open wooded habitat, particularly in the lowlands.

BLUE-THROATED BEE-EATER *Merops viridis* 27cm

Although gregarious and delight-fully aerobatic, this bee-eater appears a little less inclined to hawk its prey like other bee-eaters, preferring to wait around on exposed perches and making only short dashes to catch passing insects. Its distinguishing features are its blue throat and solid black eye-stripe and the male's rich brown cap and mantle. Its elongated central tail feathers are blue, as are the outer tail, vent and rump. The back, wings and breast, and the female's crown, are light green. Preferring drier terrain, such as open grassland or more open scrub and woodland, it occurs particularly in the coastal lowlands of Sumatra and West Java.

DOLLARBIRD *Eurystomus orientalis* 30cm

This is the only member of the roller family to occur in the region. It is usually seen in ones and twos, hunting from exposed perches in the open country; it dives on its insect prey, either in the air or on the ground. It is frequently mobbed by other birds because of its crow-like appearance. Overall, it appears dark bluish-black with an upright stance, with a large head and solid orange-red bill. In flight, two circular light blue contrasting patches on the underwing-coverts are very evident, and these give the bird its English name. Although widespread, it is not common.

Chew Yen Fook

BUSHY-CRESTED HORNBILL *Anorrhinus galeritus* 70cm

One of the smaller hornbills, this species is easily identified by its thick drooping crest. Its plumage is all black except for the greyish-brown upper two-thirds of the tail. The patch of bare skin on the throat is blue and that around the eye is either blue or white. The feet are black, as is the male's bill. The bill of the female is grey, although it can sometimes be whitish. This hornbill is gregarious outside of the breeding season, when noisy flocks of as many as fifteen birds can gather together to feed high in the tree canopy. It is found in the lowland forests of Sumatra.

WREATHED HORNBILL *Aceros undulatus* 100cm

Apart from the whitish, often dirty yellow, tail and the male's chestnut-crowned creamy head, the remaining plumage is black. The male has a bare yellow gular pouch, that of the black-headed female being blue. Like all of the hornbills, this one relies on hollow forest trees when nesting, the male restricting the size of the entrance hole with mud while the female is incubating, and the female doing likewise after hatching until the young are ready to fledge. It is particularly fond of figs, but also takes other fruits and insects. Although common in lowland and hill forests of Sumatra, its distribution in Java and Bali is confined to a few remaining undisturbed forests.

ORIENTAL PIED HORNBILL *Anthracoceros albirostris* 75cm

Recognized by its mainly black plumage with a white lower breast, belly and flanks, this species also has white patches behind the eye and at the base of the lower mandible. The huge horn-coloured bill is surmounted by a banana-shaped casque marked with black at the front end. The tail shows variable amounts of black and white, particularly on the outer feathers. It is normally found in small parties, which give a very chicken-like *puk-puk-puk-puk* cackle; also utters a loud, raucous and cackling laugh. Inhabits edges and clearings in more remote lowland forest and secondary forest right down to the coast, throughout the Greater Sundas.

Chew Yen Fook

69

RHINOCEROS HORNBILL *Buceros rhinoceros* 110cm

Although not present in Bali, this hornbill gives the impression that it is more numerous than it really is. A very large bird with a loud voice, usually seen in pairs and duetting in flight, it creates a false impression of numbers. Its head, breast, wings and back are black and the lower belly a dirty white. The tail is white with a broad black band. It can be identified in flight by its all-black underwings, white belly and banded tail. The enormous yellow bill, becoming red at the base, is surmounted by an upturned casque. Found in low densities in most large pockets of hill and lowland forest in Sumatra and Java.

Chew Yen Fook

GREAT HORNBILL *Buceros bicornis* 125cm

Usually travelling about in pairs, this huge hornbill feeds noisily in the high canopy. Its call can be described as a loud and harsh barking roar. The enormous yellow bill and casque, black face, its yellow-stained white neck and upper breast, and black subterminal bar on a white tail separate it from other hornbills of the region. The back, lower belly and wing-coverts are black, and a white wingbar is often stained with yellow. This uncommon species confines itself to lowland and submontane forests, ranging from India through South-east Asia and the Malay Peninsula to Sumatra.

FIRE-TUFTED BARBET *Psilopogon pyrolophus* 26cm

As with most barbets, the main plumage is bright green. This species has a yellow band across the breast, outlined with a black line below. The rear crown and nape are red, the chin green, the ear-coverts grey and the eye-stripe and forehead black, and it has a prominent red-tipped black hairy tuft at the base of the lores. It is very fond of ripe fruit, particularly small figs, and it will sit stationary high in the trees for long periods while frequently uttering its harsh buzzing call, which is reminiscent of a cicada. Confined to the Malay Peninsula and Sumatra, where it prefers tall forest trees in the lower hills.

LINEATED BARBET *Megalaima lineata* 29cm

This barbet has a green back, wings and tail, and a brown head and upper breast heavily streaked with buff. The remaining underparts are green and also streaked with buff. It has a broad orange-yellow orbital ring of bare skin. The heavy bill is pale pink, and the feet yellow. It is frequently seen in sparse, dry coastal woodlands, edges of savanna, acacia forest, orchards and cultivated areas. Not found on Sumatra, but quite a common resident in open forest, along forest edges and in clearings in the more remote areas of Java and Bali.

RED-CROWNED BARBET *Megalaima rafflesii* 25cm

The combination of a bright red crown, a yellow cheek patch and a blue throat is diagnostic of this medium-sized barbet. The main body plumage is bright green, and there is a small red patch below the rear of the eye which blends into black on the ear-coverts. A wide blue eyebrow separates the crown from the ear-coverts. The bill is black and the feet are grey. Juvenile birds are much duller in appearance. Like the rest of its family, it nests in holes in trees. This quite common barbet prefers the lowland forests, where it inhabits the upper canopy and is often difficult to catch sight of.

BLACK-BANDED BARBET *Megalaima javensis* 26cm

Tilford/Cooke, TC Nature

Quite large, with green body plumage and tail, this barbet has a yellow crown and a yellow spot under the eye. It also has a red throat, under which is a solid black collar which joins up to a black stripe running through the eye. The feet are pale olive-green and the bill black. As with other barbets, the feet are adapted for climbing tree trunks, having two toes pointing forward and two back. It is mainly a fruit-eater with a penchant for small ripe figs, but also eats seeds, buds and flowers. More likely to be encountered around clearings and forest edges of lowland and hill forests, this endemic of Bali and Java appears quite plentiful.

BLUE-EARED BARBET *Megalaima australis* 18cm

Morten Strange

A small barbet, more likely to be heard than seen, but careful examination of fruiting fig trees will often lead to its being located. This species' calls are a fast repeated *tuk-trrk, tuk-trrk ...*, and a series of high-pitched whistled or trilled notes. Two races occur in the region, both having green body plumage, a black malar stripe and breast stripe, and a blue chin and crown. The Javan race has patches of yellow on the cheek and below the black breast stripe. The Sumatran race lacks the yellow but has red cheeks. This is a common barbet of forests and plantations, being found from the coast to the lower hills.

COPPERSMITH BARBET *Megalaima haemacephala* 15cm

Tilford/Cooke, TC Nature

A drawn-out resonant *tonk-tonk-tonk-tonk* call, reminiscent of a vibrating hammer on metal, announces this bird's presence. While frequently found in trees in open country, it also ventures into town gardens and parks. Usually solitary, but often flocks with other barbets at fruiting trees. The back, wings and tail are green, with fawn underparts streaked with black. The race found in Java and Bali (*M. h. rosea*) has a red crown, chin, throat and cheek, the Sumatran race (*M. h. delica*) having the throat, cheek and eyebrow yellow. A thin black necklace separates the red and yellow neck markings. A fairly common resident of open lowland forests throughout the range.

LACED WOODPECKER *Picus vittatus* 30cm

Morten Strange

This woodpecker has green upperparts and nape, a yellow rump and a black tail. The underparts are buffish-yellow with the feathers edged in green, giving the laced appearance. Males have a red crown, this being black on females. The cheeks are blue-grey, bordered by a black eye-stripe and malar stripe, both of which are mottled with white. The black primaries are barred with white. The bill is black and the feet grey-green. Prefers coastal woodland and forest, and often found in mangroves, bamboo thickets and coconut plantations. It is locally common throughout the region, but is rare in Sumatra.

COMMON FLAMEBACK *Dinopium javanense* 30cm

Chew Yen Fook

Male *Female*

Usually in pairs and noticed by their continual 'churring' contact calls, the male has a red crown extending into a crest at the rear, while the female's crown is a more flattened mass of black and white feathers. Otherwise both sexes are similar, having a golden-yellow mantle and wing-coverts, red lower back and rump, black tail and primaries, and a white face with a black eye-stripe and single solid malar stripe. The white underpart feathers are edged black, forming a scaly pattern. It is unusual in having only three toes. A quite common lowland woodpecker of secondary and open forests and mangroves, but regularly appearing around cultivation, plantations and even in gardens.

WHITE-BELLIED WOODPECKER *Dryocopus javensis* 42cm

Rather solitary, this woodpecker usually makes its presence known by its loud 'laughing' and 'barking' calls, as well as its heavy hammering on dead branches. It often calls in flight with a raucous, echoing *kiow, kiow, kiow*. It has a very conspicuous appearance, being very large, with all-black plumage apart from a white belly. Males have a red crest and cheek patches. The long pointed bill is black and the feet grey. This bird of open lowland forest is found throughout the Greater Sundas as far east as Bali, but is relatively rare in both Java and Bali.

FULVOUS-BREASTED WOODPECKER
Dendrocopos macei 18cm

This smallish woodpecker is quite adaptable and fairly tame, making it relatively easy to observe. It spends a lot of time in the higher branches of tall trees, but can often be seen foraging lower down on tree trunks. Its back and tail are heavily barred black and white, and the underparts are buff and lightly streaked black. The male has a red crown and the female black. The side of the face and the chin are white, separated by a black malar stripe. It is a bird of open forest, which has a liking for plantations, parks and gardens. Although it is quite common in Bali and Java, there are few records from Sumatra.

Tilford/Cooke, TC Nature

SUNDA PYGMY WOODPECKER
Dendrocopos moluccensis 13cm

Also known as the Brown-capped Woodpecker, this little bird is usually seen on its own as it works through the dead branches and bark of trees, often close to the ground, searching out insects and larvae. Its short legs have grey feet with two toes pointing forward and two backward. Its upperparts are dark brown with white barring, giving a partially spotted appearance. The underparts are whitish with black streaking. It has a rich brown crown, black malar stripe, white face patch with a greyish-brown centre, and thin red stripe behind the eye. It prefers mangroves, open woodland and coastal secondary forest.

BLACK-AND-RED BROADBILL
Cymbirhynchus macrorhynchos 23cm

This insectivorous species is identified by its black upperparts and black band across the upper breast, its white wing-stripe, dark red underparts and throat-stripe, and its large yellow and blue bill. Juveniles have a pinkish-buff belly and grey-tinged throat. It is attracted to scrubby and wooded areas and gardens in the vicinity of water, and can frequently be found alongside rivers and streams, where it hawks insects from exposed perches. Ranges from South-east Asia through the Malay Peninsula and Borneo to Sumatra, but not found in Java or Bali. In Sumatra it is common at lowland forest edges.

BANDED PITTA *Pitta guajana* 22cm

A beautiful ground-living bird with a distinct plumage. Three races are present in the region, all having a black head with a long, broad yellow eyebrow and a white chin. The tail is blue and the back and wing-coverts brown. The wings are dark brown and have a thin white wingbar. The Sumatran race (*P. g. irena*) has breast and flanks barred blue and orange, a blue belly and orange nape, and more distinct wingbar. In West Java, *P. g. affinis* has breast and flanks barred black and yellow. *P. g. guajana* in East Java and Bali is also barred black and yellow, but has a blue bar on the upper breast. Inhabits mainly lowland forest and shaded dense thickets.

HOODED PITTA *Pitta sordida* 18cm

A forest-floor dweller, this bird hops around foraging for various insects among the leaf litter and rotting wood. It is a plump medium-sized pitta with long legs. Residents have a green plumage with a black head, blue-grey wing-coverts, and crimson undertail-coverts and lower belly. Migrants have a dark brown cap. The bill is black and the legs are pinkish. If disturbed, it flies close to the ground with rapid wingbeats. In flight, it reveals conspicuous rounded white patches on the wings. Not found in Bali, but resident and wintering northern populations occur in Sumatra and West Java.

AUSTRALASIAN BUSHLARK *Mirafra javanica* 14cm

Tilford/Cooke, TC Nature

This small bird occasionally perches in trees, but is more often seen walking around on the ground. It is a black-mottled russet-brown, with paler underparts and white outer tail feathers. Its short pinkish-grey legs have long claws. The brown bill is tinged yellow below. It has a weak undulating flight. Like many other larks, it has a fluttering hover-flight in which it often sings, with a high-pitched trilling, before gliding slowly back to ground. Not found in Sumatra, but fairly common in Java and Bali, particularly in open areas of short grass, paddyfield stubble and dry cultivated fields.

BARN SWALLOW *Hirundo rustica* 20cm

The Barn Swallow is more likely to be seen wheeling in circles and gliding with an occasional flutter high in the sky, hunting aerial insects. Depending on the weather conditions, it may come very low and be seen skimming the surface of lakes and streams. It also often perches on overhead wires and bare branches. This gregarious bird sometimes gathers in huge flocks to roost in reedbeds, tall grasses and even on city buildings. It is dark blue above, with a white breast and very long outer tail feathers. The forehead is red, and the throat red with a blue bar below. Juveniles are duller and lack the tail-streamers. A common winter visitor throughout the region, from breeding grounds in northern latitudes.

Tilford/Cooke, TC Nature

HOUSE SWALLOW *Hirundo tahitica* 14cm

Chew Yen Fook

Similar in appearance to the Barn Swallow, but lacking the long tail-streamers and the blue bar across the upper chest. The breast appears a dirty buffish-white. It otherwise has a dark blue plumage with a chestnut-red forehead. It has brown feet and a black bill. This swallow usually forms small parties with other hirundines and swiftlets to feed over water. It neither roosts nor nests communally, preferring an isolated mud nest fixed under an overhanging ledge, bridge or roof. It is resident throughout the region.

EASTERN YELLOW WAGTAIL *Motacilla tschutschensis* 18cm

Tilford/Cooke, TC Nature

Many subspecific variations in plumage occur throughout the races of this species, especially in winter plumage, and it is difficult to identify many individuals in the field. Of typical wagtail shape and behaviour, the species is olive-brown or olive-green on the back, with yellow underparts. The commonly occurring race, *M. t. simillima*, has a grey crown, yellow throat and white supercilium, and any variations could relate to other, more rare races. Immatures have browner upperparts and whiter underparts. A common winter visitor from Asia, found mainly around coastal lowlands, particularly pastureland, paddyfields and dried-up marshland.

GREY WAGTAIL *Motacilla cinerea* 19cm

Tilford/Cooke, TC Nature

A typical wagtail in shape and tail-wagging behaviour, running and skipping among the rocks and running water. Its flight is markedly undulating. It has a grey crown and mantle, with bright yellow on underparts and rump. The chin and eye-stripe are whitish and the ear-coverts greyish. The underparts of juveniles are much paler. The very long black tail is edged white. The bill is dark brown and the feet pinkish-grey. Mainly a winter migrant from northern latitudes, preferring stony river beds, streams and damp meadowland from the coast to tops of mountains.

RICHARD'S PIPIT *Anthus richardi* 18cm

Chew Yen Fook

The only pipit resident in the region. On the ground it appears as a pale, long-legged bird, often seen running around in grassland, and at other times very still in an upright posture. It is warm brown above with darker streaks, and buff below with fine streaking on the upper breast. It has a pale rump and uppertail-coverts and a prominent buff eyebrow. The flight feathers are dark brown, edged buff. In flight, the white outer feathers of the dark brown tail are very visible. Attracted to open grassland and cultivated land, dry paddyfields and roadside verges. Widespread and abundant from the coast to the mountains.

SMALL MINIVET *Pericrocotus cinnamomeus* 15cm

David Tipling, Windrush Photos

Like other minivets, this one is rather gregarious, moving among the treetops in small groups, continually calling to each other as they search out insects and ripe fruits. The head, mantle and upper back are grey, with rump, vent and outer tail orange. The long central tail feathers are black, and the black wings have a prominent yellow-orange patch. Males have a black throat and red to orange belly; females are paler, with a whitish breast. It inhabits the more open lowland forest and also trees and copses around cultivated land, gardens and mangroves. Widespread in Java and Bali.

SCARLET MINIVET *Pericrocotus flammeus* 19cm

Tilford/Cooke, TC Nature

A forest-dwelling bird which forages through the treetops in small groups. The male's head and mantle are jet-black with a blue sheen, and the underparts, rump and outer edges of the tail are red. The mainly black wing has two distinct red patches. The female has a grey crown and upperparts, with yellow to orange underparts and outer tail; the yellow underparts continue up to the chin, forehead and ear-coverts, but leaving a suggestion of a grey eye-stripe. She has two yellowish wing patches, and her rump is grey-green. Locally common throughout the region.

STRAW-HEADED BULBUL *Pycnonotus zeylanicus* 28cm

Its clear melodious song is responsible for this species' popularity as a cagebird in Indonesia, where birds are exported from Sumatra to Java and Bali. It is one of the largest bulbuls, and can be recognized by its golden-straw-coloured crown and ear-coverts and its black moustachial stripe. The back, wings and tail are shades of greenish-brown. It also has a thin black eye-stripe, and a white chin and throat, grey breast and belly, and yellow vent, the upper breast and back having a small amount of white streaking. Once a common bird of lowland and hill forest, particularly of riverine habitat, it is now becoming scarce.

Chew Yen Fook

BLACK-HEADED BULBUL *Pycnonotus atriceps* 17cm

Usually alone, but also in small groups foraging with other species, this bulbul can be recognized by its mainly yellow body plumage and glossy black head and throat. The upperparts are greenish-yellow tinged brown, and the underparts are a bit brighter. The wings are black, and the tail yellow with a black subterminal bar. A rare colour form exists in which the yellow areas are replaced by a dull grey, but it has a white vent and undertail-coverts and white tip to the tail. It is more often seen at forest edges, in riverine and coastal scrub and often along roadsides, and in secondary forest. Fairly common in lowland areas of Sumatra and Java, but much less common in Bali.

BLACK-CRESTED BULBUL *Pycnonotus melanicterus* 18cm

Sometimes referred to as the Ruby-throated Bulbul, the subspecies *P. m. dispar* is resident in Sumatra, Java and Bali and is distinguished by its bright red throat. It is quite a large olive-tinged yellow-green bulbul with black head and upright tufty crest. The underparts are paler, and the iris is either cream-coloured or dull red. It is relatively timid, keeping out of sight in well-foliaged tall trees in secondary forest and along forest edges as it searches for ripe fruits. It is reasonably common in both lowland and hill forests of Sumatra, but its numbers seem to decline going eastwards through Java.

ORANGE-SPOTTED BULBUL *Pycnonotus bimaculatus* 20cm

Ray Tipper

This is quite a noisy bird with its harsh continual toc-toc-toc-toroc call, often several birds joining in a chorus as they roam the trees in search of fruit. It is brown to olive-brown above, with paler ear-coverts and two distinguishing orange spots, one around the lores and the other above the eye. The throat is black, blending into a brown upper breast that becomes off-white with brown mottling as it leads to a white belly. The vent is bright yellow. Some birds from north Sumatra have a grey belly. A common endemic, found at montane forest edges and in clearings almost to the top of the highest mountains, where it is attracted to fruiting blueberries.

YELLOW-VENTED BULBUL *Pycnonotus goiavier* 20cm

Chew Yen Fook

This quite gregarious bird not only forages with other bulbuls but also roosts communally. It is frequently seen picking over fallen fruit and taking the occasional insect. It has a bright yellow vent, as its name implies, but is otherwise relatively sombre in plumage. The upperparts and tail are brown and the underparts white. The crown is dark brown and the face white, with black lores. It has pinkish-grey feet and a black bill. Common throughout the region, preferring a more open habitat and especially attracted to cultivation, plantations, gardens and parks, even in towns.

OLIVE-WINGED BULBUL *Pycnonotus plumosus* 20cm

Morten Strange

Like several other bulbuls, this one has rather dull, nondescript brownish-grey plumage with olive-coloured wings. It is greener and larger than the similar Cream-vented Bulbul. The throat is pale buff and the undertail-coverts pale brown. The ear-coverts are brown, streaked with white. It has red eyes, a black bill and brown legs. It keeps to the treetops as it forages for ripe fruit and insects, occasionally uttering short, soft chirping song. Found through Sumatra and West and Central Java, but not in Bali, this is a resident bird of lowland woods and forest edge, as well as secondary growth and occasionally scrubland.

CREAM-VENTED BULBUL *Pycnonotus simplex* 17cm

Morten Strange

Similar to but smaller than the Olive-winged Bulbul, this species is easily separated as the race found in Sumatra and Java has white eyes. The upperparts are a dull brownish-grey, the belly white and the vent cream. Both chin and throat are pale buff to white. The bill is black and the legs brown. An inhabitant of forest, it also ventures into more open countryside, secondary growth and sometimes plantations, but invariably it keeps to the treetops. In Sumatra it is fairly common in lowland primary forest, but in Java it is now restricted to the southern lowland forests.

RED-EYED BULBUL *Pycnonotus brunneus* 17cm

Chew Yen Fook

Another of the smaller bulbuls found in Sumatra but not in Bali or on the Javan mainland. It is very similar to but smaller and much browner than the Olive-winged Bulbul, and its ear-coverts are brown, lacking any white streaking. The underparts are buff, with the breast buff-brown. The red eyes of the adult are brown in juveniles, and the legs and bill are brown. These birds inhabit the higher canopy, where they forage for insects and ripe fruits. Relatively common throughout the lowland forests, with a tendency to be found more at the edges of the more scrubby secondary forest.

COMMON IORA *Aegithina tiphia* 14cm

This very secretive little leafbird requires patience if you want to get a good view of it. It habitually remains among the leaves of small trees, using its colour as a marvellous camouflage. It has olive-green to bright green upperparts, a white rump, and yellow underparts. The wings are black, with fairly distinct white and yellow edges and two very clear wingbars. There is a clear yellow eye-ring. Both bill and feet are bluish-grey. Mostly alone but often in pairs, it inhabits open woodland, secondary forest and mangroves, but can frequently be found in town gardens. Quite common in the lowlands of Sumatra, Java and Bali.

GREATER GREEN LEAFBIRD *Chloropsis sonnerati* 22cm

Of the seven leafbirds resident in the region, the Greater Green Leafbird is by far the largest. It is generally bright green on the back and tail, with slightly paler underparts. The male has a black throat and face with a blue malar stripe, and the shoulder of the wing is marked by a small blue spot. The female has a yellow throat edged by a blue malar stripe with a distinctive yellow eye-ring. Juveniles are very like the female with the blue malar stripe being very dull or non-existent. Typically found with other leafbirds in small flocks, high in dense foliage of tall trees in mangroves and primary or secondary forest.

BLUE-WINGED LEAFBIRD *Chloropsis cochinchinensis* 17cm

Two subspecies of this leafbird are found in the region: *C. c. icterocephala* is found in Sumatra and *C. c. nigricollis* in Java. Both have the typical colour of most leafbirds, being generally yellowish-green above and more yellow below. This species is recognized by the blue wings and edges of the tail. The male has a black patch below the eye which extends down into a yellow-bordered black throat. The Sumatran race has a yellow crown and yellow to orange nape, whereas the Javan race has a green crown and yellow upper breast. Females lack the black head and throat markings. All have a blue malar stripe, a little less evident on females. Absent from Bali.

Male (above); female (below)

GOLDEN-FRONTED LEAFBIRD *Chloropsis aurifrons* 19cm

Tilford/Cooke, TC Nature

The male is identified by the bright golden-yellow patch on the forehead and a large black throat patch extending up to the eye and across to the base of the bill. The centre of the throat is ultramarine-blue. The rest of the plumage is bright leaf-green but for a small shiny pale blue patch on the shoulder. Females are bright leaf-green all over, apart from the blue shoulder patch and a blue malar stripe. A common resident of mixed and deciduous hill forests of Sumatra. Like so many other birds with a pleasant musical song, it is trapped for the pet trade and kept as a cagebird not only in Sumatra, but in Java, Bali and worldwide.

ASIAN FAIRY BLUEBIRD *Irena puella* 25cm

Apart from the black face, neck, underparts, tail and primaries, the male's plumage is brilliant blue. Females are a duller blue with a greenish tinge, apart from their brighter blue rump. It can be distinguished from other bright blue and black birds by its larger size. More likely to be seen at fruiting fig trees, where it is often in mixed flocks with other birds. Otherwise it confines itself to foraging, usually out of sight, through the tops of tall trees of primary and secondary forest. Commonly found in the lowland forests of Sumatra, but less so in Java and not at all in Bali.

Tilford/Cooke, TC Nature

LONG-TAILED SHRIKE *Lanius shach* 25cm

The long black tail, upright stance and brown, black and white plumage make this shrike easily identifiable. It has a grey crown and nape, with chestnut mantle, back and rump; the underparts are clean white. Adults have a black mask from the forehead through the eye and across the ear-coverts. Juveniles tend to be duller, with dark grey head and nape and barring on flanks and back. Often in pairs, it roams through grassland and scrub, venturing into cultivated land and plantations up to the edge of human habitation. Quite common in open areas frequently seen perched on bare branches scanning for aerial insects or grasshoppers. Resident in Java, Bali and Sumatra.

WHITE-BROWED SHORTWING *Brachypteryx montana* 15cm

Chew Yen Fook

This tiny member of the thrush family is very timid, seldom leaving the protection of undergrowth on the forest floor. However, it becomes bolder near to mountain tops where it can often be seen in the open. It usually skulks beneath bushes, feeding on invertebrates and occasionally berries. Its presence is often advertised by its high-pitched resonating alarm call. Males have dark blue plumage all over, with a conspicuous white supercilium. Females vary being all blue in Sumatra but in Java they have a blue head and nape with rufous wings, back and tail. Ranges from Nepal through to China and SE Asia to Flores.

ORIENTAL MAGPIE ROBIN *Copsychus saularis* 20cm

Several subspecies are found in the region, including *C. s. musicus* in Sumatra and *C. s. amoenus* in East Java and Bali. The former has black wings and central tail, with white outer tail feathers, belly and vent, and a broad white patch across the wing-coverts. East Javan and Bali birds are all black, except for the white wing-covert patch. Females are much duller, with the black plumage replaced by grey. Juveniles have similar but mottled plumage. Common in the lowlands, with a wide choice of habitat from mangroves and forests to town gardens, although trapping for the pet trade has caused a serious decline.

WHITE-RUMPED SHAMA *Copsychus malabaricus* 27cm

Another beautiful songster which has declined through the actions of trappers for the pet trade. This bird spends a lot of time foraging in low dense undergrowth, hopping around on the ground. The male has a very long tail, about half its total length, with dull black central feathers and white outers, which it habitually flicks on landing. Its head, neck, back and wings are black with a blue sheen, its lower underparts are deep orange, and it has a conspicuous white patch on the rump. Common in Sumatra, where it sticks to denser parts of the forest, but far less numerous in Java.

PIED BUSHCHAT *Saxicola caprata* 13cm

Like all the chats, this bird prefers a habitat of scrubby grassland, where it perches on exposed bushes or posts to locate its insect prey. Males are all black, apart from a white rump, vent and distinctive wingbar. Females have dark brown streaked upperparts, dark brown tail, buff underparts with rufous flanks, and pale rufous-brown rump. Juveniles are spotted dark and light brown. Often sings from a prominent perch while cocking its tail. A fairly common resident of the dry open lowlands of Java and Bali, occurring sporadically in the hills; not present in Sumatra.

SUNDA WHISTLING THRUSH
Myophonus castaneus/glaucinus 25cm

Now split into two species. Chestnut-winged Whistling-thrush (*castaneus*) in Sumatra is the larger, the male having a dark blue head and upper breast, with the nape and shoulders a brighter blue; the lower back, wings and tail are dark chestnut-brown, with lighter belly and undertail-coverts. The female is duller, with blue shoulders and black crown. In the smaller Javan Whistling-thrush (*glaucinus*) the male is dark blue with blackish underparts. The plumage is more uniform than that of other whistling thrushes, lacking spangles. Both have a blackish bill. An endemic of the Greater Sundas.

BLUE WHISTLING THRUSH *Myophonus caeruleus* 32cm

Chew Yen Fook

A largish blue-black thrush having a moderate purplish sheen and sparsely flecked and spangled white on the wing-coverts. The contrasting yellow bill is often marked with black, particularly on Sumatran birds. Legs are black. It has a high-pitched screeching alarm, *creech-chit-chit chit*, and a loud whistling song often incorporating imitations of other birds' songs. A ground feeder, taking insects, invertebrates and small fruits, it is generally found in denser lowland hill forests beside rivers and among exposed limestone rock formations. Absent from Bali and not very common in Java or Sumatra.

91

ORANGE-HEADED THRUSH *Zoothera citrina* 21cm

The usual haunts of this thrush are in more secluded forest among thick undergrowth, but its high-pitched whistling alarm call betrays its presence. It has a very nice song and this, along with its attractive colouring, makes it a prime target for local bird-keepers and, more recently, the export pet trade. Its head, nape and belly are bright orange, the back grey-blue, and the tail and wings dark grey. It has a white vent and wingbar. The female has greenish-brown upperparts. Widespread from Pakistan through to the Greater Sundas, but many of those seen in Sumatra are likely migrants; in Java and Bali it is a bird of hill forest and lowland, but is not really common.

CHESTNUT-CAPPED THRUSH *Zoothera interpres* 16cm

Chew Yen Fook

Chestnut-capped as its name implies, this small thrush otherwise has a pied appearance. The chestnut crown extends down the nape, stopping at the slate-grey mantle and upperparts. It has two distinctive broad white wingbars. The throat and upper breast are dense black, leading into a black-spotted lower breast and flanks and white belly and vent. The bill is blackish-brown and the legs pink. It has a typical thrush diet of invertebrates, and often takes fruit and berries. Occurs in Java, but rarely in Bali and Sumatra.

SCALY THRUSH *Zoothera dauma* 28cm

David Tipling, Windrush Photos

In typical thrush fashion, this species spends much of the time foraging on the ground. It has only a quiet whistling song and just a feeble *tzeeet* alarm call. Its upperparts are brown with golden-brown and blackish scaly markings, these being bolder on the back and upperwing-coverts. The brown-scaled underparts are more whitish, with a pale chestnut wash on the upper breast and flanks. It prefers dense wooded areas and montane forest in Java and Bali, but in Sumatra is confined to the mountain woodland of the north. Found from north-east Europe west through to the Sundas and as far as Australia.

ISLAND THRUSH *Turdus poliocephalus* 20cm

Morten Strange

This is a medium-sized thrush with a dark grey head blending into dull blackish upperparts. The throat and upper breast are mid-grey, while the belly is chestnut and the vent white. The eye-ring, bill and legs are yellow. It has a pleasantly melodious song and a rather raucous rattling alarm call. This is a bird of the highest mountains, but it is rare in Bali. It is more common in Java, where it can usually be seen close to the volcano on Mt Tangkuban Perahu and in the mossy forests of Mt Gede. In Sumatra, the peaks of Mt Leuser are its home.

WHITE-CHESTED BABBLER *Trichastoma rostratum* 15cm

This small jungle babbler has fairly plain coloration of dark brown upperparts and buffish-white underparts. The sides of the breast are tinged grey and the flanks are tinged brown. The bill is black and the legs pinkish. The Moustached Babbler (*Malacopteron magnirostre*) and the Sooty-capped Babbler (*Malacopteron affine*) are similar in size; it can be separated from the former by not having a black malar stripe or eye-ring, and from the latter by having a brown (not black) cap and no pale eyebrow. Absent from Bali and Java, but occurs in small groups in the damp lowland forests and mangroves of Sumatra.

HORSFIELD'S BABBLER *Malacocincla sepiaria* 14cm

Chew Yen Fook

This noisy little babbler has a monotonous and repetitive call, *pee-o-eet, pee-o-eet* ..., given particularly at dawn and dusk. Usually in pairs or small groups, it forages for insects among low vegetation. The upperparts are warm russet-brown, becoming chestnut on the rump and tail. The flanks and vent are also chestnut, while the throat and belly are white and the breast grey. It has a heavy grey and black bill. Separated from Abbott's Babbler by its darker and greyer crown. This is a locally common resident throughout the region in the dense undergrowth and low thickets of hill and submontane forests.

ABBOTT'S BABBLER *Malacocincla abbotti* 16cm

Chew Yen Fook

Like Horsfield's Babbler, Abbott's is small and brown and has a heavy bill. It is distinguished by being lighter and duller and having a longer tail. The upperparts are brownish-olive, with a pale chestnut rump. The underparts are buffish-white on the throat, with a pale grey-green breast, buffish-brown belly and rufous undertail-coverts. The sides of the head are buffish-brown with a pale grey supercilium. Usually found low down at forest edges and in scrub and thickets, generally singly or in small groups. Locally common in Sumatran lowland forests, with a few records for Javan islands, but absent from Bali.

CHESTNUT-BACKED SCIMITAR BABBLER
Pomatorhinus montanus 20cm

This bird has a striking greyish-black head with a long white eyebrow, and a longish, decurved buffish-yellow bill with black at the base. The throat and breast are white, the back, flanks and vent chestnut-brown, and the tail and wings dark brown. The feet are grey. It is frequently seen in pairs or small groups with laughingthrushes, foraging in ground litter and through low bushes but also in the lower tree canopy. The subspecies from Sumatra (*P. m. occidentalis*) is quite common in lowland and hill forest, whereas birds from West Java (*P. m. montanus*) and those in East Java and Bali (*P. m. ottolanderi*) tend to occupy higher montane forests.

Tilford/Cooke, TC Nature

RUSTY-BREASTED WREN BABBLER *Napothera rufipectus* 18cm

F Lambert

One of the more substantial ground-hugging babblers, this species prefers dense cover in both lower and upper montane forest in which to seek out its mainly insect food. It can be recognized by its generally brown-streaked appearance and its white throat. The upperparts are dark brown with light brown and black streaking, and the underparts streaked dark and light chestnut. The short tail and the primaries are dark brown. Like many of the wren babblers, it has various loud whistling calls. An endemic, found particularly in the remoter mountains of the Barisan range in Sumatra.

STRIPED TIT BABBLER *Macronous gularis* 13cm

Morten Strange

The upperparts of this small tit babbler are all chestnut-brown, including the tail. The underparts, in particular the breast, are conspicuously streaked finely with black on an underwash of pale yellowish-green becoming whitish at the lower belly and vent; the Javan race is more greyish below. The sides of the head are grey, sometimes tinged yellow. The bill is brown and the legs grey-blue. A common bird in lowland Sumatra, where it lives in small groups in dense thickets, especially bamboo, and secondary growth. An occasional resident in West Java, but not found in Bali.

RUFOUS-FRONTED LAUGHINGTHRUSH
Garrulax rufifrons 27cm

Ray Tipper

This species is particularly noisy when roaming beneath the canopy with other birds, foraging for insects. It is a largish babbler with a long tail. The plumage is mostly a soft olive-brown but with a slight reddish tinge below. The throat is white to fawn, with a reddish chin and forehead, and the wings are chestnut. The bill is black and the feet a dirty yellow-brown. An endemic, inhabiting mainly primary mountain forests of the western half of Java. Although restricted in range, locally it does not appear uncommon.

WHITE-CRESTED LAUGHINGTHRUSH
Garrulax leucolophus 30cm

Tilford/Cooke, TC Nature

The white head and black face-mask with a superficial erectile crest are distinguishing marks of this cocky babbler. It is an arboreal bird, found in the lower storeys of primary and secondary hill and montane forest. Forms noisy flocks with others of its kind, foraging through the foliage with occasional visits to the ground. West Sumatra is the eastern limit of the species' natural range. The Sumatran subspecies has a dark back and short crest. The chestnut-backed subspecies found elsewhere in Asia (depicted here), is a popular cagebird in Java, and escapes are often encountered.

WHITE-BROWED SHRIKE BABBLER *Pteruthius flaviscapis* 13cm

Chew Yen Fook

Usually seen in pairs with other species, this plump little bird appears to have a proportionately large head and thickish bill. In characteristic manner it sidles along branches in search of food. Males have a black head with a conspicuous white eyebrow, grey mantle and back, and black wings and tail, with the primaries tipped white and the tertials tinged chestnut and yellow. The underparts are whitish. Females have a grey head with less distinct eyebrow, and greyish-green upperparts. Found in the more open lower montane forests of Java and Sumatra.

SILVER-EARED MESIA *Leiothrix argentauris* 18cm

This brightly coloured medium-sized babbler has a deep black head enhanced by silver ear-coverts. The mantle, breast and forehead are bright red, sometimes orange-red, as are the rump, undertail and primary coverts, and the back, tail and wing-coverts are grey-green. Both bill and legs are yellow. It has a cheerful little whistled song and a range of other chattering calls. A locally common resident of montane and secondary forests, particularly fond of dense thickets but also inhabiting forest edges and more open scrub, it occurs from the Himalayas through to Sumatra. Often kept as a cagebird.

ORIENTAL REED WARBLER *Acrocephalus orientalis* 18cm

Chew Yen Fook

This is a fairly large brown warbler with a very obvious pale buff supercilium underlined by a thin dark brown eye-stripe. Its underparts are whitish, blending into buff on the flanks, rump and upper belly, with the upper belly and the sides of the breast sparsely streaked brown. The largish bill is brown above and tinged pink below, and the legs are blue-grey. A regular winter migrant, visiting the region from its breeding grounds in eastern Asia, this species is usually found in lowland reedbeds and marshes, but frequently occurs also in rice fields and scrub close by water.

ZITTING CISTICOLA *Cisticola juncidis* 10cm

Tilford/Cooke, TC Nature

Commonly known as the Fan-tailed Warbler, this little bird occurs from Africa and Europe through India, South-east Asia and the Sundas down to northern Australia. It is a small, rather inconspicuous brown bird, heavily streaked with dark brown and buff on its upperparts. It has white underparts and vent, with warm buff flanks, and the brown and black tail is tipped white. Its white supercilium distinguishes it from the Bright-headed Cisticola. A common grassland and reedbed bird throughout the lowlands of Sumatra, Java and Bali, and naturally attracted to the wetter habitat of rice paddies.

BRIGHT-HEADED CISTICOLA *Cisticola exilis* 11cm

Morten Strange

Non-breeding birds and females are very similar to the Zitting Cisticola. They differ in having a more golden head colour and in the buff supercilium being the same colour as the nape and head sides. The throat is white and the underparts buff. The dark brown tail is tipped buff. Also known as the Golden-headed Cisticola, breeding males have a golden-yellow crown and brown rump. In the lowlands of Java and Bali this is a common bird of alang-alang grassland, as well as rice paddies and reedbeds; only locally common in Sumatra. It prefers taller grass than the Zitting Cisticola.

BAR-WINGED PRINIA *Prinia familiaris* 13cm

Morten Strange

A noisy, versatile and gregarious little bird, often uttering its loud high-pitched 'tweeting' call as it searches for insects from ground to treetop. It is recognized by its long tail of black- and white-tipped feathers rather drab olive-brown upperparts, and two distinctive white wingbars The white of the throat extends down the middle of the upper belly with the flanks pale grey and lower belly and vent pale yellow. It favours secondary growth, especially parks and gardens, mangroves, plantation and scrub. A very common endemic in the lowlands of Java and Bali, bu less so in Sumatra, particularly in the north.

YELLOW-BELLIED PRINIA *Prinia flaviventris* 13cm

This prinia has a weak song and shy habits, with a tendency to remain hidden among long grass and reeds. Appearing as a long-tailed warbler, it has olive-brown upperparts and tail, a bright yellow belly, and a very evident white chin, throat and upper breast. The head is greyish with a reddish-orange eye-ring and a very thin, and often broken, white supercilium. The legs and feet are a dull orange, and the bill is dark brown above and pale brown below. Found on Sumatra and West Java, it occurs fairly commonly in grassland, reedbeds and thick scrub in the lowlands.

MOUNTAIN TAILORBIRD *Orthotomus cuculatus* 12cm

A gregarious and timid little bird, very difficult to observe as it moves around in small groups, foraging through the thickest vegetation. It can be identified by its bell-like tinkling call of about three or four notes, culminating in a short trill. Like most of the tailorbirds, it has olive upperparts and a high-cocked tail. Apart from the orange forehead, the head and upper breast are grey, but lighter on the throat and centre of the breast, the remaining underparts and vent being bright yellow. An inhabitant of montane forests throughout the region, more commonly found in bamboo thickets, evergreen forests and scrub.

ASHY TAILORBIRD *Orthotomus ruficeps* 11cm

Tilford/Cooke, TC Nature

A small and rather plain-coloured tailorbird with a grey-brown back and greyish underparts, becoming white on the belly. Males have a dark rufous face, crown and throat, whereas females have a white throat and a much paler rufous face and crown. This energetic little bird is frequently seen in more open lowland forests, where it forages among the treetops. It also occurs in mangroves and coastal scrubland, as well as bamboo thickets and well-vegetated gardens. Common in Sumatra, but in Java found only in the northern wetland areas, and absent from Bali.

ARCTIC WARBLER *Phylloscopus borealis* 12cm

David Tipling, Windrush Photos

This small leaf warbler spends much of its time foraging through wooded areas in search of insects, often accompanying other warblers and small insectivores. It is identified by its conspicuous long yellowish-white supercilium above a blackish eye-stripe. It also has a poorly defined single white wingbar. The upperparts are dark olive and the underparts almost white, blending to olive-brown on the flanks. Similar to the smaller Yellow-browed or Inornate Warbler (*Phylloscopus inornatus*), but much duller and with less obvious wingbars. A winter visitor from northern Asia, found generally at primary and secondary lowland forest edges and mangroves.

MOUNTAIN LEAF WARBLER *Phylloscopus trivirgatus* 11cm

Chew Yen Fook

Like many of the leaf warblers, this bird is not easy to see as it flits through the tree canopy, often in mixed groups, foraging for insects. It is mainly green above, with greenish-yellow underparts. Its prominent greenish-yellow median crown-stripe and supercilium are interspaced with a blackish stripe along the crown side and a black eye-stripe. The legs are grey, and the bill black above with a reddish tinge on the lower mandible. Juveniles are duller and greener below. Fairly common in the region's montane forests, keeping to the treetops and forest edges.

ASIAN BROWN FLYCATCHER *Muscicapa dauurica* 12cm

Chew Yen Fook

This species of submontane forest and forest edges can frequently be found in plantations, open forest and even gardens. It hunts for aerial insects from exposed perches, often shaking its tail on arrival back at the perch. It is greyish-brown above and whitish below, becoming grey-brown on the flanks and the sides of the breast, and has a pale eye-ring. The black bill has a yellow base to the lower mandible. Almost certainly a winter migrant to the region from north-east Asia, but some may be resident in north Sumatra. Occurring throughout the lowlands of the Greater Sundas, it is particularly common on offshore islands.

VERDITER FLYCATCHER *Eumyias thalassinus* 16cm

This is a fairly large blue flycatcher. In typical flycatcher fashion it chooses exposed perches from which it can launch its attacks on passing aerial insects. It prefers open forest and forest edges, but is often seen perched high in the tree canopy. The male's plumage is a fairly uniform greenish-blue, with white fringing to the undertail-coverts giving a scaled appearance. The area around the lores is dark grey to black. Females look duller and greener, and juveniles are a mottled brown tinged with green. The feet and bill are black. Found in the lowlands and hills of Sumatra.

Tilford/Cooke, TC Nature

INDIGO FLYCATCHER *Eumyias indigo* 14cm

Morten Strange

The plumage of this species is mainly a deep indigo-blue colour especially on the upperparts and breast, the blue becoming dense blue-black on the throat and face towards the base of the bill. The forehead is tinged white, leading into a bluish-white supercilium. The belly is grey becoming white and then buff at the vent. The legs and bill are black. Confined to dense forest, where it often mixes with other species low in the canopy and often close to the ground. It is a relatively common resident of montane and submontane forests in Java, but less so in Sumatra, and is not found in Bali.

LITTLE PIED FLYCATCHER *Ficedula westermanni* 11cm

Chew Yen Fook

This species is sexually dimorphic, the male being black and white and the female brown and white. The male's upperparts are all black, apart from a white wingbar and supercilium and also white edges to the base of the outer tail feathers. The underparts are all white, and the legs and bill black. Females have brown upperparts and whitish underparts, the tail being rufous-brown. It frequently mixes with other species, feeding throughout the forest, and it has a quiet *pi-pi-pi-pi* call followed by a vibrating churring sound. Relatively common at the edges of submontane forests above 1000m.

HILL BLUE FLYCATCHER *Cyornis banyumas* 15cm

Males have dark blue upperparts with iridescence on the forehead, eyebrow and shoulder. The chin, lores and a thin line around the eye are black, as is the bill, while the throat and breast are rufous-orange, blending into white at the lower belly and vent. The legs are dark brown. Females have brown upperparts with an off-white eye-ring, and underparts similar to the male's but paler. This is a bird of high mountain forests, favouring low undergrowth and bamboo thickets and the edges of forest glades. Ranges from Nepal through China to Borneo and Java, but it is not found in Bali or Sumatra.

MANGROVE BLUE FLYCATCHER *Cyornis rufigastra* 15cm

Not only are different races of the Mangrove Blue Flycatcher present in the region, other very similar species also occur which can cause confusion. Considering all factors together, it is not difficult to separate this one. It has blue upperparts and orange underparts, although the female is paler and has a whitish chin and a white loral patch. It is most likely to be confused with the Hill Blue Flycatcher (*Cyornis banyumas*), which also has a clear melodious, warbling song, but that has a pale blue forehead and blackish chin. Usually inhabits coastal forests, plantations and mangroves close to the beach.

GOLDEN-BELLIED GERYGONE *Gerygone sulphurea* 9cm

Chew Yen Fook

Also known as the Flyeater, this is the only Australian fairy-warbler found in the region. As a tiny but active bird, and well disguised among foliage, it is noticed more by its shrill fluty song. Its upperparts are brown-tinged grey, and its underparts bright yellow. The chin and neck are white, and it often shows a whitish loral spot. The dark grey brown tail is tipped black, with a subterminal line of white spots. It shows a preference for open forest and bamboo and conifer thickets, as well as mangroves and even plantations, and is quite common up to around 1500m.

BLACK-NAPED MONARCH *Hypothymis azurea* 16cm

When this beautiful flycatcher leaves the shade of woodland and comes into the open, its delightful azure-blue head and back are very obvious. The male's head has a short erect black crest on the hindcrown and a thin black band above and below the bill. Another narrow black band extends across the upper breast, with the grey belly paling to a white vent. The wings are grey and the tail blue-black. The bluish bill has a black tip. Females have a duller blue head and greyer breast, without any of the male's black markings. It prefers lowland and secondary forests, where it is usually common.

PIED FANTAIL *Rhipidura javanica* 19cm

This very active small bird is a typical fantail in its wing-drooping posture and regular flicking and fanning of the tail. The plumage is black and white, and it has a diagnostic black band across the otherwise white underparts. When the tail is fanned, the broad white tips to the feathers are very obvious. The upperparts are a dusky black, and there is a thin white eyebrow. Both bill and legs are black. It is found particularly in mangroves, open woodland and secondary growth at elevations of up to 1500m, and is often present in gardens. A common resident throughout the region.

GOLDEN WHISTLER *Pachycephala pectoralis* 17cm

Morten Strange

The male of this species has quite brilliant colouring of black, white and bright yellow. The yellow nape patch is joined by a narrow yellow band right through to the breast, belly and vent, while the head is black except for a white chin and throat, leaving a narrow black band between the throat and belly. The upperparts are olive-green and the tail is almost black. Females are much duller, having drab olive upperparts and dirty buff underparts, with the lower belly and undertail-coverts tinged yellow. Found in hill and montane forest in East Java and Bali, but absent from Sumatra.

GREAT TIT *Parus major ambiguus/cinereus* 13cm

Tilford/Cooke, TC Nature

Of the world's 33 subspecies of Great Tit, two are resident in the region, both very similar in appearance but quite different from those at the extremes of the species' range. Those in Sumatra, Java and Bali are relatively small, black, white and grey birds. The head and throat are black, with a black strip from throat down to vent; the ventral strip is wide on males, but on females can be very narrow and even broken. There is a large white patch on the cheek and a small white nape spot. Usually in pairs or small groups foraging for insects, particularly in mangroves and open forests, and often comes to gardens.

VELVET-FRONTED NUTHATCH *Sitta frontalis* 12cm

Tilford/Cooke, TC Nature

A very active and attractive bird which, unlike other nuthatches in the region, is well adapted to creeping both up and down tree trunks in search of insects and spiders. It is identified by its red bill and reddish feet, dirty pinkish-white underparts and whitish throat. The upperparts are mainly violet-blue, with black-edged feathers in the tail and primaries, and the forehead is velvety-black. Usually seen in pairs but occasionally in small parties, moving erratically through the understorey branches of forests and plantations. Fairly common in lowland and hill forests in Sumatra and Java.

BLUE NUTHATCH *Sitta azurea* 13cm

Chew Yen Fook

This nuthatch typically forages along branches and tree trunks, levering off fragments of bark to get at grubs and insects. In shady forest it appears to have black upperparts, but in fact its back, wings and tail are shiny dark blue; only the crown, nape and sides of the face are black. The throat and breast are buffish-white. East Javan birds (*S. a. azurea*) have a bluish-black lower belly and vent, but these are black on West Javan and Sumatran birds (*S. a. nigriventer* and *S. a. expectata*). It has a yellow bill and grey feet. Not recorded in Bali, but in Sumatra and Java fairly common in lower mountain forests.

CRIMSON-BREASTED FLOWERPECKER
Prionochilus percussus 10cm

Chew Yen Fook

The adult male is brilliantly coloured, with bright yellow underparts being broken only by a bright crimson breast patch. The upperparts are greyish-blue, with a bright red patch on the crown. The forehead and primaries are black and the tail is dark blue-grey. It has a white malar stripe finely underlined in black. Females are olive-green above, with only a tinge of orange on the crown, and have a yellow-tinged grey throat, a white malar stripe and a yellow belly with greyish-olive flanks; the undertail-coverts are distinctively white. A rather scarce and local bird of Sumatran and West Javan lowlands.

YELLOW-VENTED FLOWERPECKER
Dicaeum chrysorrheum 9cm

Chew Yen Fook

With its diagnostic plain bright yellow or orange undertail-coverts and its black-streaked white remaining underparts, this bird is unmistakable. Juveniles are duller, having only a pale yellow vent and greyish underparts with less obvious darker grey streaking. It has a repeated contact call, *tzit-tzit-tzit*, uttered especially in flight. This little flowerpecker seeks out its diet of insects and small fruits in open forest and secondary growth, occasionally venturing into gardens. It is found throughout the region's lowlands, but nowhere is it common.

ORANGE-BELLIED FLOWERPECKER
Dicaeum trigonostigma 9cm

This tiny little bird frequents the tops of small trees, searching out small insects as well as small ripe fruits. It is very active, flitting among the forest canopy, garden trees and mangroves throughout the lowlands of the region. The orange belly, vent, rump and lower back, and the blue-grey head, wings and tail are diagnostic of the male. The female is pale olive-grey with a grey throat, becoming yellowish on the belly, the rump being dirty orange. The bill is black and the legs grey. Juveniles look like very dull females lacking yellow or orange.

PLAIN FLOWERPECKER *Dicaeum concolor* 8cm

Chew Yen Fook

Very small and rather drab compared with the males of other flower-peckers, this bird is easily confused with females and juveniles of similar species and so may be overlooked. Both sexes have olive-green upperparts and greyish-olive underparts, becoming whitish-yellow along the centre of the belly up to the throat. Yellowish-white pectoral tufts are occasionally visible. It has a black bill, finer than that of many other flowerpeckers, and dark grey-blue legs. It inhabits hill forest and secondary growth, as well as plantations and cultivated areas, and is attracted to mistletoe. Found throughout the region.

111

SCARLET-BACKED FLOWERPECKER *Dicaeum cruentatum* 9cm

Males show scarlet-red on the crown, back and rump, and black on the sides of the face and neck as well as on the wings and tail. The underparts are greyish, with a wide pale buff band down the centre of the belly and on the throat. Females are olive-brown with paler throat and belly, have a scarlet rump and uppertail-coverts, and lack the reddish-washed crown and mantle of the Scarlet-headed Flowerpecker (*Dicaeum trochileum*). A fairly common resident of Sumatran lowland and submontane forests, scrub, plantations and gardens, even in cities, it is also attracted to mistletoe.

BROWN-THROATED SUNBIRD *Anthreptes malacensis* 13cm

Chew Yen Fook

Of the region's 12 different sunbirds, only Brown-throated and Olive-backed occur in Bali. They are the most common, found throughout the region, and are usually seen in open plantations, gardens, coastal scrub and mangroves. Identified by their bright coloration and long curved bill, they usually hover in front of flowers in search of nectar. The male Brown-throated has a yellow breast, belly and vent, and a brown throat often fringed with dark purple. The face and chin are olive and the upperparts are iridescent olive-green, blending into a dark bluish head with a green sheen. The female is olive-green above and yellowish below.

RUBY-CHEEKED SUNBIRD *Chalcoparia singalensis* 10cm

Morten Strange

Although this small sunbird is brightly coloured, its iridescence when in the shade is not often visible and males even appear blackish. In bright light, however, the male's iridescent dark green upperparts and crown are brilliant and the ear-coverts reveal a deep red. The throat is orange-brown, blending into a yellow belly. Females (depicted here) are duller, with olive-green upperparts and pale orange and yellow underparts. The bill and legs are black, the legs often with a greenish tinge. This bird of forest edge and sparse woodland occurs alone and in pairs, often in the company of other species.

WHITE-FLANKED SUNBIRD *Aethopyga eximia* 13cm

This colourful sunbird is aptly named after the cluster of soft white feathers on its flanks. Apart from the rather drab olive and black wings, longish blue-green tail and olive underparts, the male has a bright yellow rump and red throat and upper breast, the last crossed by a bluish-green necklace-like band, and his crown is iridescent purple. Except for the white flanks, the shorter-tailed female is dull olive, a little lighter on the throat and vent. This species is endemic to Java, where it is resident in mountain forests at forest edges and alpine scrub, and is particularly fond of flowering trees.

COPPER-THROATED SUNBIRD *Leptocoma calcostetha* 13cm

Chew Yen Fook

Males dark, even appearing blackish. Upperparts have green iridescence and the breast and malar stripe purple. The throat and upper breast show a dark coppery-coloured gleam, while the flanks are yellow and the tail bluish-black. Separated from the smaller Purple-throated Sunbird by lacking the red breast but having yellow flanks. Females have a greyish head, dark olive-green to brown upperparts and black tail; their undertail-coverts and throat are pale grey, blending into a greenish-yellow breast. A mostly lowland resident in Java and Sumatra, found in coastal scrub, woodland and mangroves.

OLIVE-BACKED SUNBIRD *Cinnyris jugularis* 10cm

A very active and often noisy little bird, either flitting between flowers or hovering to extract nectar. It also feeds partly on insects and pollen. Quite small, with a long, curved black bill, it has olive-green upperparts with darker wings and black tail. The underparts are bright yellow, with white showing below the tail. Breeding males have a black chin and upper breast with a purple iridescence. The nest, a wonderfully woven structure of fine grasses and hair-like materials, is suspended precariously from the end of a branch or among foliage. It is found throughout the region, particularly in lowland areas.

Chew Yen Fook

TEMMINCK'S SUNBIRD *Aethopyga temminckii*
male 13cm, female 10cm

This bright red sunbird has a relatively long and slightly notched red tail. Its upperparts, throat and breast are crimson-red, apart from the iridescent purple eyebrow which extends around the nape and the malar stripe. The uppertail coverts are also purple adjacent to a bright yellow rump patch. It has greyish-white underparts. Females are smaller with a shorter reddishbrown tail. They are dark olive-green above, with greyish head, wings tinged brown, and paler and yellowish below. It is an occasional bird of the more open montane forests of Sumatra and Borneo.

Chew Yen Fook

JAVAN SUNBIRD *Aethopyga mystacalis* 12cm

Now considered a separate species endemic to Java, this little sunbird, along with Temminck's Sunbird (*Aethopyga temminckii*), was previously considered to be a subspecies of the Crimson Sunbird (*Aethopyga siparaja*). It also goes by the name of Violet-tailed Sunbird. The male has a bright crimson head, breast and back, yellow rump, and long iridescent purple tail. The crown and malar stripe are also purplish, the wings olive-green and the underparts whitish. The female, very much smaller and with only a short tail, is dull olive-grey. It inhabits forests and forest edges, usually in pairs, and is particularly attracted to mistletoe flowers.

LITTLE SPIDERHUNTER *Arachnothera longirostra* 15cm

An inconspicuous little olive and yellow bird with a long, curved bill which is black above and grey below. The upperparts are drab olive-green and the underparts brilliant yellow. It can be separated from all similar species by its whitish throat. It inhabits gardens, plantations and logged forests, where it can often be seen moving rapidly across open areas as it searches for nectar, in particular from the flowers of banana and ginger. Common throughout the region, especially in lowland and hill forests, and in some areas way into the mountains.

LONG-BILLED SPIDERHUNTER *Arachnothera robusta* 21cm

Chew Yen Fook

This very long-billed and largish spiderhunter is normally a solitary bird, even to the extent of driving away all other spiderhunters that encroach on its territory. It has a habit of perching high on exposed branches and uttering its repeated and uninteresting call, *chew-lewt, chew-lewt*. In flight, it gives a simple *chit-chit* call. Its upperparts are olive and its underparts yellow, with the breast and throat streaked with olive-green. Separated from the Spectacled Spiderhunter (*Arachnothera flavigaster*) and the Yellow-eared Spiderhunter by lacking yellow ear patches or eye-rings. A resident of the lower hill forests of Java and Sumatra.

YELLOW-EARED SPIDERHUNTER
Arachnothera chrysogenys 17cm

The Yellow-eared is another sombrely plumaged spiderhunter which has olive-green upperparts and is olive-grey below, with moderate streaking on the breast. The vent is yellow. It has a largish yellow cheek patch and narrow and often incomplete eye-ring. It is not found in Bali, but in Sumatra it may be confused with the Spectacled Spiderhunter (*Arachnothera flavigaster*), which is, however, a more robust bird with a shorter bill and broad yellow eye-ring. It is usually found in lowland forest, along forest edges and in gardens.

GREY-BREASTED SPIDERHUNTER *Arachnothera affinis* 17cm

Chew Yen Fook

More likely to be encountered around banana thickets and high in the tops of flowering trees, this species' most significant feature is the long, decurved bill. Its plumage is fairly dull, being olive-green above, with fine black streaking on the crown, and olive-tinged grey below with rather obscure black streaks. Tail and wings tend to be blacker. The subspecies in Java and Bali (*A. a. affinis*) has a pale yellow front edge to the wing and a yellow tinge to the breast. Sumatran birds (*A. a. concolor*) have a duller, more washed-out appearance. A common resident of coastal and lowland woodland and thickets of the region.

ORIENTAL WHITE-EYE *Zosterops palpebrosus* 11cm

An active bird with upperparts bright olive-green, and a bright yellow throat merging into a greyish abdomen and undertail-coverts; wings and tail are darker. It has a ring of bright white feathering around the eye contrasting with darker face sides, and sometimes a yellowish forehead. The lores are black, bill dark brown and feet grey. Often in large flocks with other small birds, roaming the canopy. Ranges throughout the Sundas, inhabiting forest edges and mangroves; common in lowlands and hills up to about 1500m. A popular cagebird in Indonesia.

MOUNTAIN WHITE-EYE *Zosterops montanus* 11cm

More likely to be observed in noisy parties as it forages for insects through the treetops. Similar in size to other small white-eyes, this darker species resembles the Black-capped White-eye (*Zosterops atricapilla*) in Sumatra, but without the black on the head. Its upperparts are olive-green, with slightly darker wings and tail. The underparts are buffish-grey, becoming brownish on the flanks and yellow on the vent. The throat is also yellow. A very common bird of the region's lower mountains, preferring open forest and forest edge as well as plantations and copses in agricultural land. It is seldom found in the lowlands.

JAVAN WHITE-EYE *Zosterops flavus* 10cm

Smaller and paler than most other white-eyes, this species has fairly uniform bright olive-yellow upperparts and plain bright yellow underparts. Legs and bill are black, and it has a brown iris. It should not be confused with the slightly larger Lemon-bellied White-eye, which is a little duller and has a black loral spot. More a bird of coastal scrub and thickets, as well as mangroves, it is found in small flocks with other white-eyes, feeding on insects and nectar. A fairly rare endemic of Java's north-east coast, but more common in south Borneo.

RED AVADAVAT *Amandava amandava* 10cm

Tilford/Cooke, TC Nature

For most of the year males and females look alike but, as the breeding season approaches, sexually dimorphic plumage becomes very apparent. Females have greyish-brown upperparts and greyer sides to the head, with greyish-buff underparts and blackish wings and tail. The rump and bill are bright red. Breeding males become crimson-red on the back and breast, with small white spots on breast sides and flanks and also on the rump. The feet are pinkish-grey. Although still fairly common in East Java and Bali, trapping has reduced its numbers drastically in West Java and it thought extinct in Sumatra.

PIN-TAILED PARROTFINCH *Erythrura prasina* 15cm

Tilford/Cooke, TC Nature

This pretty finch often mixes with large flocks of munias, raiding rice crops as they become ripe. The upperparts are green, the rump and tail red, and the lower belly and vent brownish-buff. The male has elongated central tail feathers, a red patch in the centre of the belly, and blue throat, cheeks and forehead. The female's head is greenish. A bird of scrubland and thicket, particularly bamboo, adjacent to cultivated land and rice paddies. Resident in Java and Sumatra but not in Bali, it was once very common but is now confined mainly to the lowlands.

JAVAN MUNIA *Lonchura leucogastroides* 11cm

Chew Yen Fook

This munia has dark chestnut-brown upperparts, tail and vent, black chin, throat and sides of face, and a white belly. It should not be confused with the White-bellied Munia (*Lonchura leucogastra*) which has fine pale streaking on the upperparts, brown flanks and yellowish-brown tail. Normally a grassland bird, it is attracted to the easy pickings of cultivated land and especially rice paddies at harvest time. At other times it congregates in small groups, noisily foraging through grassland. An endemic to the region and particularly common in Java and Bali; in Sumatra, it is found mainly in the east.

SCALY-BREASTED MUNIA *Lonchura punctulata* 11cm

Tilford/Cooke, TC Nature

Familiar to the cagebird enthusiast as the Spicebird or Nutmeg Mannikin, this species is easily identified by the scaly appearance of its breast and flanks, the white feathers being edged with brown. The upperparts and throat are brown. Like other munias, it is often regarded as a pest to the farmer, raiding paddyfields and cultivated crops, but it also inhabits secondary scrub, gardens and open grassland. A very common munia, found from the coast up to the lower mountain slopes, and regularly exploited as a cagebird. Its range covers India through to Sulawesi.

WHITE-HEADED MUNIA *Lonchura maja* 11cm

Tilford/Cooke, TC Nature

The White-headed Munia is very much like other munias, gathering outside the breeding season in mixed flocks to exploit seeding grasses and especially the rice harvest. It is quite small, with the head, nape and throat completely white. The remainder of the plumage is chestnut-brown, apart from a small area of black on the undertail-coverts. The bill and feet are pale bluish-grey. It prefers open grassland, as well as swamps, reedbeds and rice paddies, and is fairly common throughout the region. It is often kept and exported as a cagebird.

JAVA SPARROW *Padda oryzivora* 16cm

Adults are largely grey, with a black head and throat, large white cheek patches, a black tail and white undertail-coverts, and a heavy pink or red bill. Juveniles are brownish-grey, with a buff or brown breast. It is usually found in small groups feeding on rice and seeding grasses, and in other low vegetation, mangroves and scrub. Seen in cities, towns and villages, but far less commonly than in the past. A endemic of Java and Bali, it has been widely introduced over much of Southeast Asia. This once common bird has declined alarmingly in recent years, probably due to agricultural practices and the demands of the pet trade.

Tilford/Cooke, TC Nature

EURASIAN TREE SPARROW *Passer montanus* 14cm

This is perhaps the commonest of all resident species throughout the region. It has a chestnut crown, black throat, sides of face and ear patch, pale buff underparts, brown back and wings mottled with black and white, and two pale wingbars. It is commonly seen feeding on the ground, and in large flocks raiding seeding crops. It is found particularly in open habitats with scrub and low vegetation, from the coast to mountain villages. It is well distributed, particularly in the lowlands, and has colonized almost all areas, from cities to forest clearings. Its range extends from Europe eastwards to Australia.

Tilford/Cooke, TC Nature

STREAKED WEAVER *Ploceus manyar* 14cm

Breeding males of this colonial species have a black head with golden-yellow cap, dark brown upperparts with russet-brown feather edges giving a streaked appearance, and white underparts with black streaking on the breast. Females and non-breeding males have a black-streaked brown head and buff eyebrow and chin. Large nesting colonies are established in isolated trees near good feeding areas, the polygamous males leaving the females alone to weave their elaborate suspended nests. After breeding they form large roaming flocks, stopping off at seeding grassland and often around rice paddies. They occur locally around reedbeds, rice paddies and swampy grasslands in the lowlands of Java and Bali.

Morten Strange

BAYA WEAVER *Ploceus philippinus* 15cm

In breeding plumage, the male has a bright yellow crown and nape and black cheeks and throat, the upperparts being streaked and mottled dark and light brown and the underparts buff. Females have the crown streaked and mottled brown, a whitish chin, a buff eyebrow, and light brown cheeks and upper breast. This master nest-builder suspends its tightly woven grass nest from the leaf of a palm tree or convenient branch, the tunnel entrance below the globular nest body ensuring protection from rain and predators. Numerous nests are concentrated on isolated trees. Found mainly in open areas of the lowlands and hills of Sumatra and West Java, it has become rather scarce in East Java and Bali.

Morten Strange

ASIAN GLOSSY STARLING *Aplonis panayensis* 20cm

Adults are all-black and can be separated from the Short-tailed Starling by being larger and having a green iridescence. Juveniles have black-streaked whitish underparts, with upperparts streaked black and brown. Both have dark red eyes. Feeds mainly on soft ripe fruits hanging on the trees, but also takes insects, rarely visiting the ground. Often congregates in big noisy flocks to roost and also nests colonially. A common lowland bird of open areas near forest and secondary vegetation and coastal scrub, but also attracted to coconut plantations and cultivated areas and gardens, frequently in towns and cities. It is found throughout the region.

Tilford/Cooke, TC Nature

ASIAN PIED STARLING *Gracupia contra* 24cm

Tilford/Cooke, TC Nature

This bird's gregarious nature is evident in its communal night-time roosting habits and its daytime feeding behaviour, when small flocks probe and search the ground for insects and other invertebrates. Like many starlings, it is noisy and utters the typical squawks and whistles of its family. The upperparts are mainly black, tinged brown. It has a white wingbar, with forehead, cheeks, belly, vent and rump also white. Bare orange skin around the eye, yellow feet and white-tipped orange bill give the bird a colourful appearance. Common in Java, Bali and south Sumatra, especially in more open and cultivated lowland country.

BLACK-WINGED STARLING *Acridotheres melanopterus* 23cm

This striking starling is all white, apart from black wings and tail and a yellow patch of bare skin around the eye. The tail is tipped white and there is a white wingbar. Three subspecies occur which differ in the amount of black in the plumage, birds in the west of the region becoming much lighter and having black only on flight feathers and tail. It lives in pairs or small groups and is attracted to short grassland and even mown lawns. Endemic to Java, Bali and Lombok, but often seen as a cagebird throughout the region. It is in fact a protected species, and is now thought to be declining in number. However, it may still be found in the open lowlands, and can be seen in some towns and gardens in East Java and Bali.

Tilford/Cooke, TC Nature

BALI MYNA *Leucopsar rothschildi* 25cm

This beautiful myna is endemic to Bali. Its distinctive plumage is entirely white, apart from black wing-tips and the tip of the tail. It has a long white crest, and a patch of sky-blue skin around the eye. It prefers the drier lowlands of West Bali. This is one of the world's most threatened species, with very few birds now surviving in the wild. The destruction of habitat and collecting for the pet trade are the main causes of its decline, but efforts are being made worldwide to secure its future by increasing the protection it receives and by establishing a captive-breeding and release programme.

HILL MYNA *Gracula religiosa* 30cm

Another species under great pressure from habitat loss and trapping for the pet trade, this bird is particularly in demand for its ability to mimic, as well as for its own enormous repertoire of whistles and calls. It is black overall, with orange wattles both behind and below the eye, and has a white wingbar. It has yellow feet and an orange bill. Although it occasionally congregates in small flocks, it is more likely to be seen in pairs in tall trees. Once a common bird of lowland forest edge, it has become rare in Bali and Java but remains in reasonable numbers in Sumatra.

BLACK-NAPED ORIOLE *Oriolus chinensis* 26cm

The male is a brilliant yellow, with a distinct black stripe passing through the eye and across the nape. The flight feathers and tail are black, edged yellow. The female is much duller. Juveniles show olive where adults show black, and they have black-streaked whitish underparts. Keeps to the tree canopy while foraging for insects and soft fruit. A fairly common bird throughout the region's lowlands and hills, favouring secondary and open forest, parks and gardens, mangroves and coastal scrub. Widely kept as a cagebird throughout the region for its melodious fluting calls and brilliant colouring.

BLACK-AND-CRIMSON ORIOLE *Oriolus cruentus* 22cm

Chew Yen Fook

This is a bird of the treetops, rarely seen on the ground. It often occurs in pairs, but frequently alone, as it forages for insects and ripened fruit. It has a cat-like mewing call. Easily identified as a medium-sized oriole, the males having all-black plumage apart from the crimson-red upper belly and a small crimson patch on the primary coverts. It has a distinctive pale blue-grey bill and blue-black legs. Females are all black, with the breast tinged chestnut and streaked with grey on the belly. Found mainly in the hill forests of Sumatra, but less so in Java and not at all in Bali.

BLACK DRONGO *Dicrurus macrocercus* 28cm

Unlike most other drongos, this one favours open country and often associates with cattle, which disturb insect prey. It is frequently seen in open scrub, cultivated land and rice paddies, where it perches rather conspicuously on fencing, isolated branches and telephone wires alongside roads. It has dull black plumage, with a long, deeply forked tail with outer feathers which tend to turn outwards at the tips. Juveniles show pale grey barring on the underparts. A common resident of Java and Bali, but the few birds found in Sumatra are thought to be migrants.

ASHY DRONGO *Dicrurus leucophaeus* 29cm

Chew Yen Fook

Usually observed in pairs waiting on exposed branches to swoop on large passing insects, this drongo has the typical long, forked tail and upright stance. Its plumage is light blue-grey, with a tuft of black feathers at the base of the upper mandible. There are five subspecies within the region, all differing slightly in their grey coloration. Mimics other birds, and has mewing-type calls as well as a loud song. A common bird of open woodland and forest edge and clearings, resident throughout the region's hills and mountains.

LESSER RACKET-TAILED DRONGO *Dicrurus remifer* 26cm

This spectacular drongo, fairly big and glossy black, has a square-ended tail with extended outer tail-streamers which are about 50cm long. Each streamer consists of a feather shaft which terminates with an elongated oval web around 10cm long. A substantial tuft of short black feathers covers the base of the upper mandible, creating a long-headed appearance. It is smaller than the Greater Racket-tailed Drongo, has no frontal crest, and has no fork in the tail. Inhabits dense montane forest and secondary forest, especially along the edges, in Sumatra and West Java.

HAIR-CRESTED DRONGO *Dicrurus hottentottus* 32cm

Two aspects of this drongo's plumage are rather unusual. It has a peculiarly fork-shaped tail, which has the outer feathers spread outwards and upwards almost into a lyre-shape. It also has an extraordinary crest of long hair-like feathers from the front of the crown, hence its name, but this feature varies among races. The general plumage is glossy black with an iridescent spangling, particularly on crown and upper breast. In East Java and Bali it is common in open areas of lowland and submontane forests. In Sumatra it is replaced by the Sumatran Drongo. It hawks insects from a low perch.

Chew Yen Fook

GREATER RACKET-TAILED DRONGO *Dicrurus paradiseus*
30cm plus 35cm tail rackets

A large drongo with a forked tail, and with beautifully extended outer tail feathers terminating in rackets on the outer edge of the feather shafts, the rackets twisted in a short spiral. Its forked tail easily separates it from the Lesser Racket-tailed Drongo. The plumage is a glossy black, and adults often display a short frontal crest on the crown. The bill and legs are black. It is usually found in pairs, often hawking for insects in glades in swamp, primary and also secondary forests and in mangroves. It remains a common bird of the lowland forests of Sumatra, but it is becoming rarer in Java and Bali through habitat loss.

Chew Yen Fook

WHITE-BREASTED WOODSWALLOW
Artamus leucorynchus 18cm

This is the only resident woodswallow in the region. It shows similarities to the true swallows in its stance and its gliding flight, but it can easily be distinguished by its more stocky appearance, broad triangular-shaped wings and squarish unforked tail. Its quite heavy bill is grey, and the entire upperparts except for the white rump are dark slate-grey. The underparts are white. This species feeds on insects, which it catches in flight. Quite a common bird of open spaces, particularly in the lowlands and hills. Frequently seen on posts and wires.

Tilford/Cooke, TC Nature

CRESTED JAY *Platylophus galericulatus* 28cm

This is a relatively inconspicuous bird until it starts to call, when its rather harsh and rattling sounds soon reveal its presence. Its plumage is very dark grey to black, apart from a broad white neck patch. It is easily distinguished by its tall, flat, upright crest. Its bill is black and the legs dark grey. It is frequently observed in small, very noisy groups foraging through trees and bushes in search of large insects. Found in lowland forest in Java and Sumatra, but not in Bali.

RACKET-TAILED TREEPIE *Crypsirina temia*
35cm including 18cm tail

This member of the crow family has a uniformly blackish plumage with a greenish-bronze sheen. It has black feet, a strong black bill and blue eyes. It differs from all similar birds in the region in having a long tail with spatulate ends to the central feathers. Usually seen alone or in pairs, it inhabits lowland and hill forests and plantations, often occurring in secondary growth and the more cultivated areas, as well as scrubland and gardens. It is resident in both Java and Bali, where its numbers are diminishing owing to habitat destruction and collecting for the pet trade.

LARGE-BILLED CROW *Corvus macrorhynchos* 48cm

Chew Yen Fook

By far the largest crow of the region. Usually in pairs or small flocks, scavenging where it can, it is nevertheless a wily creature and keeps its distance from man. It is distinguished by its large size and very heavy bill. It is black all over with blue and green iridescence. The area of the forehead appears rather small compared with the size of the bill. In flight, its broad wings with spread primaries are beaten slowly and it often calls a harsh *kaar* as it goes. Fairly common in open country, around villages and at all altitudes.

FURTHER READING

Howard, R. and A.A. Moore. *Complete Checklist of the Birds of the World*. 2nd edition. Oxford University Press, Oxford, 1991

Inskipp, T., N. Lindsey and W. Duckworth. *An Annotated Checklist of the Birds of the Oriental Region*. Oriental Bird Club, Sandy, England, 1996

Jepson, P. and R. Ounstead. *Birding Indonesia – A Birdwatcher's Guide to the World's Largest Archipelago*. Periplus Editions, 1997

King, B., M. Woodcock and E.C. Dickinson. *A Field Guide to the Birds of South-East Asia*. HarperCollins, London, 1975

MacKinnon, J. and K. Phillipps. *A Field Guide to the Birds of Borneo, Sumatra, Java and Bali*. Oxford University Press, Oxford, England, 1993

Mason, V. and F. Jarvis. *Birds of Bali*. Periplus Editions, 1989

Sibley, C.G. and B.L. Monroe, Jr. *Distribution and Taxonomy of Birds of the World*. Yale University Press, New Haven and London, 1990

Steffee, N.D. *Field Checklist of the Birds of Java and Bali*. Russ Mason's Natural History Tours, Kissimmee, Florida, 1981

White, T. *A Field Guide to the Bird Songs of South-east Asia*. National Sound Archive, London, 1984 (tape and notes)

Whitten, A.J., D. Sengli, A. Jazanul and H. Nazarudin. *The Ecology of Sumatra*. Yogyakarta Gadja Mada University Press, 1987

Whitten, T. Roehayat Emon Soeriaatmadja and Suraya A. Afiff. *Ecology of Indonesia Series Vol. II. The Ecology of Java and Bali*. Periplus Editions, 1996

Voous, K.H. *The Breeding Seasons of Birds in Indonesia*. Ibis 92 (1950): 279–287

USEFUL WEBSITES

Kukila – the journal of Indonesian ornithology: www.kukila.org

World Wildlife Fund – information on the charity's work in Sumatra and Borneo: https://www.worldwildlife.org/places/borneo-and-sumatra

Birdlife International Indonesia Programme: http://www.birdlife.org/asia/partners/indonesia-burung-indonesia

INDEX